Enzymatic Fuel Cells
Materials and Applications

Edited by

**Inamuddin[1,2,3], Mohammad Faraz Ahmer[4],
Mohd Imran Ahamed[5] and Abdullah M. Asiri[1,2]**

[1]Centre of Excellence for Advanced Materials Research, King Abdulaziz University,
Jeddah 21589, Saudi Arabia

[2]Chemistry Department, Faculty of Science, King Abdulaziz University,
Jeddah 21589, Saudi Arabia

[3]Department of Applied Chemistry, Faculty of Engineering and Technology,
Aligarh Muslim University, Aligarh-202 002, India

[4]Department of Electrical Engineering, Mewat College of Engineering and Technology,
Mewat-122103, India

[5]Department of Chemistry, Faculty of Science, Aligarh Muslim University,
Aligarh-202 002, India

Copyright © 2019 by the authors

Published by **Materials Research Forum LLC**
Millersville, PA 17551, USA

Published as part of the book series
Materials Research Foundations
Volume 44 (2019)
ISSN 2471-8890 (Print)
ISSN 2471-8904 (Online)

Print ISBN 978-1-64490-006-2
ePDF ISBN 978-1-64490-007-9

Distributed worldwide by

Materials Research Forum LLC
105 Springdale Lane
Millersville, PA 17551
USA
http://www.mrforum.com

Manufactured in the United States of America
10 9 8 7 6 5 4 3 2 1

Table of Contents

Preface

Enzymatic biofuel cells are a kind of bioelectronic device that uses oxidoreductase enzymes as an electrocatalyst to catalyze the transformation of chemical energy into electrical energy. However, conventional fuel cells use precious metal nanoparticles as the electrocatalyst. The applications of conventional fuel cells have wide dimensions of their use in stationary, transportation and in some portable devices. The conventional fuel cells in spite of major applications have the limitations to apply in implantable gadgets as well as some portable applications due to biocompatibility and reusability issues. Interestingly, enzymatic biofuel cells have various positive qualities for energy conversion, including renewable catalysts, the flexibility of fuels, and capacity to work at room temperature. The enzymatic biofuel cells in contrast to conventional energy systems use enzymes as catalysts to encourage the conversion of chemical energy into electrical energy. These enzymes are also able to catalyze fuels such as sucrose, fructose and glucose. In addition to their use as catalysts, they are biocompatible in nature. Due to this fact, the theme of enzymatic biofuel cells has broadened during the last several years, in fact, it is currently easy to apply in many applications including implantable gadgets, for example, biosensors, pacemakers, catheters, defibrillators, dynamic conveyance gadgets (for instance, insulin pumps), tranquillize conveyance frameworks, cochlear implants, self-controlled artificial muscles, and remote detecting and some specialized gadgets in electronics. These wonderful highlights and serious applications in different fields have inspired scientists in creating biofuel cells.

The **"Enzymatic Fuel Cells: Materials and Applications"** book is intended to explore various aspects of biofuel cells including fuel cell electrochemistry, use of enzyme and enzyme immobilization techniques, use of materials such as mesoporous materials, graphene composites, conducting polymer composites and applications of biofuel cells.

We are appreciative to all the contributing authors and their co-authors for their nice chapters. We may like to thank all publishers and authors who had given permission to use their figures, tables, and schemes. Be that as it may, great effort has been made to get the copyright endorsements from individual proprietors and referenced to the imitated materials; we may need to offer our sincere regrets to any copyright holder if unexpectedly their advantage is being infringed on.

Inamuddin[1,2,3], Mohammad Faraz Ahmer[4], Mohd Imran Ahamed[5] and Abdullah M. Asiri[1,2]
[1]Centre of Excellence for Advanced Materials Research, King Abdulaziz University, Jeddah 21589, Saudi Arabia

[2]Chemistry Department, Faculty of Science, King Abdulaziz University, Jeddah 21589, Saudi Arabia

[3]Department of Applied Chemistry, Faculty of Engineering and Technology, Aligarh Muslim University, Aligarh-202 002, India

[4]Department of Electrical Engineering, Mewat College of Engineering and Technology, Mewat-122103, India

[5]Department of Chemistry, Faculty of Science, Aligarh Muslim University, Aligarh-202 002, India

Enzymatic Fuel Cells
Materials Research Foundations **44** (2019) 1-28

Materials Research Forum LLC
doi: http://dx.doi.org/10.21741/9781644900079-1

Chapter 1

Methods of Enzyme Immobilization on Various Supports

Amna Ahmad[1], Muhammad Rizwan Javed[1], Muhammad Ibrahim[2], Arfaa Sajid[3],
Khadim Hussain[1], Muhammad Kaleem[1], Hafiza Mubasher Fatima[1], Habibullah Nadeem*[1]

[1]Department of Bioinformatics and Biotechnology, Government College University, Faisalabad, Pakistan

[2]Department of Applied Chemistry, Government College University, Faisalabad, Pakistan

[3]Department of Chemistry, University of Lahore, Lahore, Pakistan

* habibullah@gcuf.edu.pk

Abstract

Enzyme immobilization has become an essential process in the industry, medicine, and biotechnology over the last ten years. Scientists have developed many techniques which contain methods varying from physical adsorption and covalent attachment to entrapment in polymers and sol-gels. Immobilization of enzyme on cellulose nanofibers, nanoparticles and carbon nanotubes for fabrication of biofuel and biosensors and for the synthesis of biocatalysts are emerging as an innovative research area. This chapter will provide an overview of the recent development in enzyme immobilization techniques.

Keywords

Adsorption, Covalent Bonding, Cross Linking, Entrapment

Contents

1.   Introduction

Enzymes which are also called biochemical catalysts catalyze a large number of reactions. Universally enzymes exist in animals and plants. These biocatalysts have been extensively employed in different areas because of their substrate specificity, green chemical nature and ease of production. They have been widely used in many applications like starch conversion, dairy products, beverage processing e.g., wine, beer, fruits, vegetable juices and baking processes [1,2,3,4,5]. They have gained industrial importance because of their influence on the end products in the textile industry [6]. The implementation of enzymes has become an essential processing strategy in industries like detergents, as well as in paper and pulp manufacturing [7,8]. Due to the catalytic nature of enzymes, concern regarding medical and chemical preparation have been increased in pharmaceuticals, chemical and food industry [9,10]. Enzymes are also used in ravage organization, particularly for hard dissipates management and effluent sanitization [11,12,13]. Previously enzymes were also used for the conversion of biomass to biofuels like bioethanol, biodiesel, biogas and biohydrogen [14].

Enzymes are important biomolecules and are accountable for all those inter-conversion processes which are needed to maintain life [15]. About 4000 biochemical reactions are known to be catalyzed by enzymes [16]. Nobel laureate Emil Fischer recommended in 1894 that enzymes were very precise and both the enzyme and the substrate have particular complementary geometrical structures that fitted appropriately into one another. Frequently, this is also called "the lock and key" model.

Many enzymes are larger in size than the substrates on which they act and just a small fraction of enzyme is directly involved in the catalysis. The area containing these catalytic residues binds the substrate and then completes the reaction is known as the active site. Few enzymes also have binding sites for tiny molecules which are commonly direct or indirect substrate or product of the catalyzed reaction. Such binding can help to enhance or reduce the activity of enzymes, giving a source for feedback regulation. Similar to all proteins, enzymes are extensive, linear chains of amino acids that fold up to generate a three-dimensional product.

In most considerate environmental and experimental situations, enzymes being catalysts have some brilliant characteristics e.g., selectivity, specificity and high activity which allow executing the complex chemical processes [17,18,19]. Therefore, from biological entities, the engineering of enzymes to industrial reactors is a very exhilarating aspiration. The enzyme immobilization has been a very powerful tool to enhance almost all characteristics of enzymes for example activity, stability, selectivity, a decrease in inhibition and specificity. In order to immobilize the enzymes, the employment of immobilization solution which allows controlling the enzyme and supporting interaction through different orientations and different conditions is the main factor which is responsible to increase the chances of success [19].

By definition, immobilization means the synthesis of something which is fixed or immobile. In 1971 the first enzyme engineering conference held by Henniker, NH, USA defined that immobilized biocatalysts or enzymes are physically permanent in a distinct area in order to catalyze a particular reaction with frequent use and without any loss of catalytic activity [20]. In other words, immobilization can also be described as a key to optimize operational performance of biocatalysts in industrial processes [21,22].

Conversely, their biological resources make enzymes inappropriate from industrial aspects: commonly enzymes are soluble, repressed by substrates, products and other components, with average stability and with no perfect catalytic properties vs. non-physiological substrates [23]. To resolve the problems of protein stability, enzyme immobilization is the easiest method. The method of immobilization also enhances the control of the reaction, keeps away the contamination of product via enzymes and allows

Enzymatic Fuel Cells
Materials Research Foundations **44** (2019) 1-28

Materials Research Forum LLC
doi: http://dx.doi.org/10.21741/9781644900079-1

the use of diverse reaction patterns [24]. On the other hand, an adequate immobilization system can also be a better solution for many additional enzyme restrictions. The combination of immobilization with stabilization is one of the most common objectives of immobilization [25,26]. Though a controlled immobilization can become not just a solution for enhancing the stability of enzymes, but also decreases the inhibition by increasing selectivity and specificity of the enzymes [19,27].

The physical and chemical characteristics of enzymes could be changed by immobilization which might cause difficulties in the separation. The most significant modifications observed are the enzyme stability, steric and conformational effects, kinetic features, diffusional and transfer effects [28]. There are two important functions exhibited by immobilized enzymes:

(i) The non-catalytic function which provides an easy separation enabling easy control of process and reuse of catalysts.

(ii) The catalytic function which changes the substrate into the desired product.

The major advantage of enzyme immobilization is the use of the same enzyme continuously. This makes the whole system more suitable by prevention of enzyme loss at the end of every batch and also helps in the designing of the continuous system. While for biocatalysts, the immobilization causes more stable activity, it is easy to control the process. The extra use of substrate could be avoided because in immobilization the ratio of enzyme and substrate is very high [29]. The enzyme can be easily eliminated from the reaction at the end of the process which removes possible risks for a bioprocess coming from contamination [30].

The system of immobilization stops the inhibition of the product and permits us to work at an increased concentration of substrate. This process, on the whole, is very beneficial from an economic point of view because it enhances productivity as well as the stability of the enzyme and decreases unfavorable effects. Other principle advantages provided by immobilization are securing space, controlling process inefficient manner, providing more consistent quality and exhibiting acceptable expenses [31]. In view of all these applications which immobilization tag along, the most important benefit would be entire cost effectiveness as it possibly the most essential point of attraction in the industry.

2. Enzyme immobilization

In 1916 the first immobilized enzyme was discovered [32]. It was shown that invertase activity was not vulnerable during absorption on a solid matrix (aluminium hydroxide or charcoal). This point directed to the progress of recent accessible technique of enzyme immobilization. Primarily, the immobilization technique accustomed to having very small enzyme loading, regarding accessible surface areas. Several methods regarding covalent enzyme control or immobilization were developed in the late 90s. However, immobilization of enzymes has been investigated for a long period but the emergence of currently published literature showed a constant attraction in this area [33,34]. Being very efficient, the commercial advantages of immobilized enzymes have increased [3,34]. Additionally, during enzyme immobilization on solid support, its resistance to several environmental alterations like temperature or pH has been enhanced [35]. In contrast to its free form, enzymes in the immobilized situation are usually very consistent and easy to operate. Besides, in food and pharmaceutical industries products are functional as they remain sterilized at the end of the process. Furthermore, in a matter of protease, upon immobilization the speed of autolysis procedure can be radically declined, if a multisubunit and multi-point immobilization is attained or the condition of enzyme favorable environment is achieved [36].

Furthermore, a fixed enzyme develops a lot of characteristics like pH tolerance, performance in organic solvents, heat resistance and functional stability or selectivity. Enhancing the stabilization of multimeric enzymes and structural rigidity of the protein avoids dissociation regarding inactivation [37,38]. Without the requirement of frequent time consuming, purification procedure and costly extraction, the immobilized enzyme is again set for successive reactions. These modifications cause structural changes into the enzyme molecule through the implemented procedure of immobilization and unlike bulk, solution create a microenvironment in which enzyme can perform their function [39]. The major objective of enzyme restriction is to increase the benefit of enzyme catalysis that can be feasible via implementing a support with elevated binding capacity along with a reduction in the cost of synthesis [40].

The stability of natural or non-restricted enzyme is primarily resolute through its native composition while constancy of restricted enzyme is greatly based on numerous aspects, involving binding position, nature of its relation with carrier, the liberty of the conformational modification inside matrix, microenvironment in which enzyme is placed, carrier physical or chemical formation, characteristics of spacer (e.g., neutral, charged, hydrophobic or hydrophilic, length, bulk) connecting enzyme molecule to carrier. Thus their stability regarding temperature, duration and further situations of storage and experimental variables might be expected to have an effect on immobilization [19]. It has been observed that in comparison of natural enzymes most of the enzymes immobilized through diverse techniques of immobilization have enhanced activity. For instance, absorption of epoxy hydrolase on diethyl amino ethyl cellulose (DEAE) via ionic attachment was two times more dynamic in contrast to natural enzyme [41]; lipase-

Enzymatic Fuel Cells   Materials Research Forum LLC
Materials Research Foundations **44** (2019) 1-28   doi: http://dx.doi.org/10.21741/9781644900079-1

lipid composite captured in n-vinyl-2-pyrroildone gel matrix was 50 times more vigorous than natural a catalyst [42].

Maintenance of activity via carrier bound immobilized enzyme is commonly just about fifty per cent. At high loading of enzymes, diffusion restriction might happen because of their imbalanced delivery in a permeable carrier directing to the decrease of clear activity [43]. The conditions for increased maintenance of activity are marginal, therefore needing a lengthy screening of immobilization conditions like pH, loading of enzyme, binding chemistry and carrier. Alteration in characteristics of enzymes does not essentially signifies developments and in some examples, a careful and easy procedure should be employed to maintain better features of using enzymes together.

Immobilization of enzymes may be done through diverse methods; generally characterized as chemical and physical methods. The physical process has weak interaction among enzymes and matrix while in chemical processes there is the creation of a covalent bond between enzyme and the support. In recent years, the progress regarding the restriction of site-selective protein has noteworthy development. Progress in organic chemistry and molecular biology has directed to the improvement of various very strong, effective and site-specific applications of attaching protein onto supports. All this happened via improvement of well-designed protein microarrays, continuous flow reactor and biosensor [44,45].

There are two significant roles of an immobilized enzyme i.e.

(i) Management of process and reprocessing of catalyst for a simple separation of non-catalytic function of the enzyme

(ii) The catalytic function of the immobilized enzyme is important to convert the substrate into the ideal product.

It is necessary to make an adequate technique of immobilization to immobilize the biocatalyst from which the catalytic and non-catalytic requirements can be executed. The emerging technique that contains stabilizing of an enzyme in or on the insoluble matrix is the enzyme immobilization. The system of immobilization aids the whole process because of having many applications; however, these systems have few problems which could delay the process as there are no definite techniques to be implemented for every enzyme in the industry [22].

3.   Techniques of enzyme immobilization

Because of the fact that chemical nature and reactive groups of the binding site of the enzyme stop the loss of enzyme activity, the choice of the enzyme immobilization system

Materials Research Forum LLC
doi: http://dx.doi.org/10.21741/9781644900079-1

is incredibly significant. Noteworthy information related to the nature of the active site can be useful. Furthermore, the active site can be secured through binding of protective groups and can be later on eliminated without any failure in enzyme activity. Such a secured function is feasibly satisfied through a substrate or viable inhibitor of the enzyme in some situations. Many general procedures of enzyme restriction include covalent pairing, adsorption, crosslinking and entrapment. Each process is briefly discussed below [33,46].

3.1 Adsorption

In a comparison of other methods of immobilization, adsorption of enzymes onto insoluble supports is an old and easy process with extensive applications of enzyme loading (Fig.1). The enzyme can be immobilized through simple mixing of the enzyme with an appropriate adsorbent under favorable situations of pH and ionic power. After the removal of slackly attached or unattached enzymes, the adsorbed enzymes are secured by the contact with the hydrophobic interfaces, proteolysis and agglomeration [47]. The selection of adsorbent is especially based on reducing the escape of the enzyme used. For the prevention of chemical changes and damage to the enzyme, concern is required for accessible surface characteristics of catalysts and support. Adsorption by a physical process usually moves in the direction of significant alteration within the protein microenvironment and normally contains multipoint protein adsorption among a single molecule and many binding sites of the protein [48]. The major drawback relating this procedure is that the enzyme can be desorbed by variation in temperature, substrate and ionic concentration.

Fig. 1 Immobilization of enzymes by adsorption

3.2 Covalent bonding

It involves the creation of covalent bonds between enzymes and the support matrix (Fig. 2). Functional groups of enzymes get connected to the matrix since for the catalytic performance of these efficient groups are not accountable. The binding reaction is performed under circumstances that do not result in failure of activity and the dynamic sites of the enzyme remains unchanged by the reagents utilized. The covalent linkage

Materials Research Forum LLC
doi: http://dx.doi.org/10.21741/9781644900079-1

between enzyme and support occurred due to side chain amino acids (i.e., aspartic acid, arginine, histidine) and their level of reactivity depends on diverse efficient groups such as indolyl, phenolic hydroxyl, limidazole [49].

Fig.2 Immobilization of enzymes through covalent binding

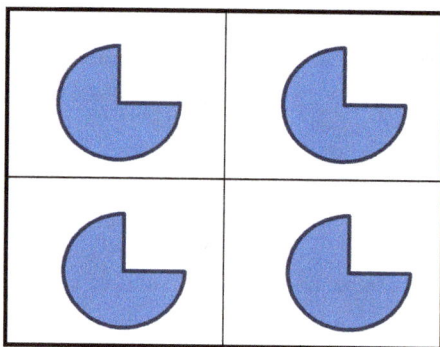

Fig.3 Entrapment of enzymes

When used for enzyme association, peptide modified surfaces ended in high activity or constancy with restricted protein orientation [50]. Often, support material activated the functional groups via definite reagents and enzyme and is then attached to the support material by covalent association. Cyanogen bromide activated-sepharose, CNBr-agrose involving carbohydrates moiety and glutaraldehyde like spacer arm have passed on the thermal stability to covalently bound enzymes [51,52]. Linkage among the carrier and enzyme could be acquired via both undeviating associations in a component or through an intercalated connection of conflicting length, which is named as a spacer. These linkages provide a higher level of mobility to attach biocatalysts thus their activities can be improved in comparison of the directly attached biocatalyst.

3.3   Entrapment

Enzyme entrapment can be expressed as the limited movement of enzymes in porous gel, however keeping them as a free molecule in the solution. The process of enzyme entrapment in gels or fibers is a suitable method for use in processes including small molecular weight substrates and products. Though, a problem which huge molecules in the impending catalytic site of entrapped enzymes prevented the use of captured enzymes by means of increased molecular weight substrate. The method of entrapment includes a covalent attachment. Enzymes have been captivated in native polymers such as agarose, gelatine and agar by thermo reverse polymerization however in alginate and carrageenan through ionotropic gelation. Many synthetic polymers have also been used such as polyvinyl alcohol hydrogel and polyacrylamide [53,54].

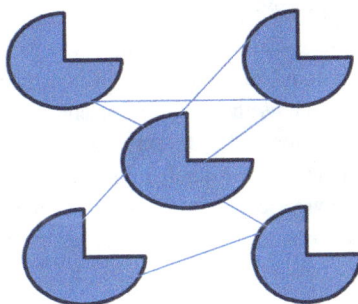

Fig.4 Cross linking of enzymes

3.4   Crosslinking

This process (Fig. 4) includes binding of biocatalysts to each other via bifunctional or multifunctional ligands. The cross-linking method is very easy and simple but not a desired process of immobilization because cross-linking does not implement any support matrix. Since the cross-linking process includes a covalent type bond formation, the biocatalyst immobilized in this manner often undergoes alterations in structure through a consequential failure inactivity. For cross-linking, glutaraldehyde has been the mostly used bifunctional agent, because of its reduced cost, great effectiveness and constancy. The reactive aldehyde groups at the two ends of glutaraldehyde react with a free amino group of enzymes. The cells or enzymes have been commonly cross linked in the existence of an immobile protein such as gelatin, collagen or albumin and can be

Materials Research Forum LLC
doi: http://dx.doi.org/10.21741/9781644900079-1

implemented to either cells or enzymes. The major drawback is the unwanted activity loss which can emerge from the contribution of catalytic groups.

4. Current advances in the immobilization techniques of enzymes

4.1 Smart immobilization by magnetic nanoparticles

Smart immobilization has been thought as an innovative immobilization method due to efficient covalent bonding of polymers, the presence of magnetic constituent and the stimulus receptive effect. These smart polymers undergo important conformational changes during reaction due to environment alterations for example pH, temperature and ionic strength [55]. Furthermore, magnetic elements can be removed from reaction mixture via utilization of magnetic decantation [56]. In the development of innovative catalytic systems, important advancements have been made which involve immobilization on magnetic nano-carriers. In comparison to the magnetite, nanoscale units of the magnet have substantial superparamagnetism, immense surface vicinity, the enormous ratio of surface to volume, thus capable of enhancing loading capability and decreasing the dispersion restriction [57]. One more major advantage of nanoparticles involving magnetic nanoparticles is maintaining the perfect direction of enzymes on supports. In comparison to polymeric supports, nonporous magnetic nanoparticles have no peripheral diffusion difficulties enabling them further competitive, particularly for high range industrial practice in solid or liquid systems [58].

Enzyme immobilization is beneficial for commercial applications because of the ease in operating, the convenience of enzymes separation from the reaction mixture and reprocesses therefore, making it cost effective for use in processes e.g., constantly or fixed bed operation [59]. Furthermore, a fixed enzyme without difficulty penetrates the skin, thus has little or no allergenicity. An additional advantage is commonly improved immovability in both operational conditions and storage for example in direction of denaturation through heat or through autolysis [45].

Despite related benefits of the immobilized enzyme onto magnetic nanoparticles, it also has few drawbacks such as conformational alteration of the enzyme, lowered activity, modification in characteristics, mass transfer restrictions, reduced efficacy and probability of denaturation of enzyme against the insoluble substrate. Furthermore, exploitation of immobilized enzyme on an industrial scale for long duration needs the synthesis of very firm derivatives having appropriate functional characteristics for a certain reaction i.e., selectivity and activity [60].

Materials Research Forum LLC
doi: http://dx.doi.org/10.21741/9781644900079-1

4.1.1 Applications of immobilized enzymes on nanoparticles

Nanoparticles are widely used for the immobilization of important industrial enzymes. Table-1 lists the applications of immobilized enzymes.

Table-1: Applications of immobilized enzymes

Enzyme	Nanoparticles	Applications	References
α-Amylase	Silica nanoparticles	Formula of detergents for improving elimination of starch	[61]
Aspergillus niger β-Glucosidase	Iron oxide Nanoparticles	Production of biofuels	[62]
Superoxide dismutase	Nano Fe_2O_3 coated On surface of gold electrode	In development of biosensors	[63]
Lysozome	Chitosan nanofibers	For antibacterial	[64]
Tyrpsin	Nanodiamond synthesized via ignition	In process of proteolysis	[64]
Candida rugosa and *Pseudomonas cepacia* lipases	Zirconia nanoparticles	Declaration of ibuprofen and phenylethanol,	[65]
Mucor javanicus lipase	Nano-sized magnetite	Solvent free production of diacylglycerols	[66]
Horseradish peroxidase	Magnetite silica nanoparticles	In Immunoassays process	[67]
Cholesterol oxidase	Nano Fe_3O_4 coated	investigation of entire cholesterol in serum	[68]
Thermoanaerobacter brockii alcohol dehydrogenase	Gold and silver Nanoparticles	Production of alcohol	[69]
Haloalkane dehalogenase	Silica layered iron oxide nanoparticle	Proteins blend having dehalogenase sequences	[70]
Lipase	Polystyrene Nanoparticle	Esterification and aminolysis	[71]

Enzymatic Fuel Cells   Materials Research Forum LLC
Materials Research Foundations **44** (2019) 1-28   doi: http://dx.doi.org/10.21741/9781644900079-1

4.2   Site-specific enzyme immobilization onto *E. coli* curli nanofibers

Lately, an innovative surface of bacteria like amloid nanofibers of biofilms has been investigated [72,73]. Stacked to each other as well as to surface or interface, biofilms are matrices encapsulating the bacteria [74]. With constituents of additional extracellular matrices, these supra-molecular polymers secured the cells from hazardous metals, chemicals and physical pressures, enabling biofilm dependent materials most appropriate for the applications in industry [75,76].

A protein immobilization platform has been created by scientists that changes curli nanofibers, the amyloid fiber element of *E.Coli* biofilms and protein entrapped covalently with a peptide domain [73]. In this method, bio-film integrated nano fiber display (BIND), heterologous functional peptide domains are genetically combined to CsgA (curli fiber subunit protein in *E.coli*) amyloidogenic protein. In curli fiber when CsgA peptide fusions were brought together, the peptide domains turn in the functional handle that can be utilized to change characteristics of fibers, to entrap metals, template nanoparticles growth or to improve linkage to the surface. Current progress in the engineering of this method has revealed that the curli pathway may be implemented to transfer a range of CsgA functional chimaeras and also entirely heterologous amyloidogenic sequences, recommending that this can be the most common method to produce well-designed material [77,78]. In the research named as catalytic BIND; it was shown that a highly applicable industrial enzyme α-amylase can be fixed on curli fibers of *E.coli* bio-film. A genetically programmable irreversible immobilization approach was used, spontaneous formation of covalent bond between 13-AA SpyTag and 15 kDa SpyCatcher (SpyTag/SpyCatcher is a suitable pairing tool for irrevocable peptide-peptide ligation) split protein takes place [79]. The SpyTag combined with the CsgA accumulated into fibers that are highly similar to the natural curli fibers with SpyTag available for coupling to SpyCatcher. When SpyCatcher is combined with the α-amylase, the immobilization reaction became very vigorous, along with the capability to produce site-specific binding among two components even in a complicated mixture. The activity of enzyme and immobilization on bio-film were differentiated by the implementation of a filter plate assay and presented that α-amylase cell activity is interrupted. The results imply that this technique might be capable to fuse the scalability of entire cell catalysis with the modularity of enzyme surface immobilization via transformation of *E.coli* bio-film extracellular matrices into designer functionalized surfaces [80].

Materials Research Forum LLC
doi: http://dx.doi.org/10.21741/9781644900079-1

4.3 Organic-inorganic hybrid nanoflowers: A new approach in enzyme immobilization

A number of diverse nanosized and macro materials have been implemented for the immobilization of enzymes to enhance the stability and catalytic activity of the enzyme. Usually, immobilized enzymes with traditional technology of immobilization have better stability whereas their activity is less as compared to the free or natural enzyme. Lately, a well-designed technique of immobilization was explored in the production of flower type organic-inorganic hybrid nanostructures with astonishing stability and catalytic activity. In this innovative approach of immobilization, protein or enzyme and metal ions behaved as organic and inorganic components respectively, in order to produce hybrid nanoflowers. It was confirmed that the hybrid nanoflowers greatly improved catalytic activities and stability in an extensive variety of experimental conditions e.g., temperature, salt concentration and pH in comparison to natural and conserved immobilized enzymes [81].

This one step and easy immobilization depend on the coordination reactions among amino groups in the enzyme backbone and nanocrystals of copper phosphate. The nanoflower type structure configuration completed in three successive steps, nucleation and formation level of primary crystals is the first step, next is the stage of crystals and the last one is the formation stage of nanoflower. The process of improvement of enzyme activity and stability can be illustrated with these impending aspects: the first reason is the high surface area of nano scale captured enzyme, the second is the lowered restrictions of mass transfer and the third is the supportive effect of the nano-scale captured enzyme and the last aspect is the suitable enzyme molecules conformation in the nano-flowers. Because of these extraordinary characteristics, it was considered that these hybrid flower type structures can be implemented in an extensive range of scientific and technical processes [81].

4.4 Immobilization of laccase on titanium nanoparticles for degradation of micropollutants

In wastewater, the micro-pollutants can be efficiently destroyed by an enzymatic treatment. But, its application in industry is inhibited because of the synthesis of the expensive purified enzyme. This research presented an innovative technique to directly fix or immobilize the domestic crude enzyme extracts from *Pleurotus ostreatus* (oyster mushroom) on to the functionalized titanium or TiO_2 nanoparticles' surface. Complete research was performed to comprehend the relation among complicated crude enzyme extracts and the immobilization support. Through simple dilutions of the crude extract of the enzyme, the efficiency of immobilization can be highly increased. In comparison to immobilized commercial enzymes, the consequential biocatalytic nano-particles had analogous performance. At last, the micropollutants destruction potential of the biocatalytic nano-particles was confirmed by destroying two micro-pollutants usually noticed in sewage called as bisphenol-A and carbamazepine. The effective laccase extraction and immobilization on biocatalytic nano particles has been a cost efficient substitute for the treatment of conserved waste water containing intractable micro-pollutants [82].

The immobilization of enzyme is desirable for the advanced operational stability in practical application under several industrial situations. It also assists in the design of biocatalytic reactors with simple enzyme reuse and control of process [83]. Lately, the titanium dependent enzyme immobilization technology has attracted attention, because of its reduced cost, better constancy, management capability with amine and carboxyl groups and suitable biocompatibility. In previous investigations, successively functionalized titanium nanoparticles with 3-aminopropyltriethoxylane and glutaraldehyde were reported to binding covalently with purified commercial laccase [84,85]. The immobilized laccase maintained most of its genuine activity and the bio-catalytic nanoparticles were found to have better operational stability for the destruction of micropollutants [82].

4.5   *In-situ* and post-synthesis immobilization on nanocrystalline metal-organic framework

Metal-organic frameworks (MOFs) are permeable substances, composed of metal ions or clusters and organic linkers. Their growing applications during the last few years verify the high capability of these innovative materials [86]. Currently, metal-organic frameworks have been treated as host matrix of enzyme restriction [87,88]. The most instantaneous method has to create metal organic frameworks with pores sufficiently huge to host enzymes [89,90]. Though, this method is distant from being worldwide since it involves two important difficulties:

(i) Only some MOF materials acquire enough mesopores

(ii) Limited to enzymes having miniature molecular dimension, capable to be placed in these holes.

However, currently, some investigations have provided the chances of creating enzyme-metal-organic framework biocatalyst without hosting enzymes in the structural holes of metal organic frames materials, by post synthesis [91,92] or *in-situ* methods [93,94]. Furthermore, the creation of the enzyme-metal-organic framework complexes can now take benefit of the astonishing structural and compositional adaptability in addition to hierarchical porosity of metal-organic framework material because of the use of an extensive range of organic functional moieties, nature of metal, size of particle and manufacturing conditions etc.

In an investigation, post-synthesis and *in-situ* approaches were compared with immobilization provided huge sized enzyme β-glucosidase in two MOF components at room temperature and in the water like an inimitable solvent [95,96]. In normal situations, the immobilized enzymes secured little portion of their activity. The effects of diverse constraints like intercrystalline mesoporosity of metal-organic-framework support; its organic performance and the systematic alteration in manufacturing conditions, in case of *the in-situ* method have been investigated and conferred. Experiments with other enzymes and diverse MOF supports were also performed.

The methods reported up to now including metal-organic framework like enzyme supports revealed that

(i) MOF substances and enzymes mixture encouraged uncertain types of relationships among enzymes and previously made MOF

(ii) The rapid and restricted increase of MOF crystals were capable to accidentally entrap a definite quantity of enzyme molecules [97]

(iii) The enzymes encouraged the creation of MOF dependent coating [98]

(iv) The enzyme as a portion of the metal-organic framework behaved like a node [99,100]

(v) The safety of the resultant enzyme MOF complex utilizing polymers such as poly-dopamine to improve hydrophilicity and biocompatibility of support and for enhancement in stability and reusability of biocatalyst [101,102].

A method depending on entrapping the enzymes into the comparatively arranged mesopores synthesized throughout the agglomeration or aggregation of MOF nanocrystal has been reported [103]. In contrast to other approaches, this *in-situ* method did not produce implanted enzymes in the metal-organic framework crystals although they were probably placed in a vacant space among the nanocrystals. Instead of documented post-synthesis methods, chances to entrap enzymes in crystalline mesoporosity have been added to effortless relation among enzyme and exterior surface of metal-organic framework dependent support. In the case of the *in-situ* immobilization method, chances of manufacturing MOF at room temperature and in water were carried out like initial point so as to protect enzymatic activity. The preparation in a non-aqueous system using N, N-dimethylformamide and its influence on the functionality of free β-glucosidase and metal organic framework enzyme complex was investigated. The colloidal nature of the

Materials Research Forum LLC
Materials Research Foundations 44 (2019) 1-28   doi: http://dx.doi.org/10.21741/9781644900079-1

synthesis media may ignore the leakage of the enzyme from the support, and thus providing extremely high competence of *in-situ* immobilization method [103].

4.6 Enzyme immobilization on cellulose matrix

Among viable biomaterials, cellulose has been widely investigated by theoretical and empirical methods because of its biodegradability, chemical stabilization, biocompatibility and reduced threats to the atmosphere [104]. Cellulose is mainly a profuse biopolymer on the planet, present in hemp, wood, cotton and further plant dependent materials and serves as the main strengthening stage in plant structure [105]. Cellulose is also prepared by tunicates, algae and few bacteria [106]. Derivatives of cellulose like carboxymethyl cellulose, cellulose acetate and cellulose nitrate are commercially useful substances in biological as well as chemical industries because of reduced cost, non-hazardous, renewable, biodegradable and biocompatible [107]. Significantly, derivatives of cellulose have efficient groups to bind with enzymes. Derived forms of cellulose are perfect substrates of enzyme immobilization [108]. Several technologies have been implemented in immobilizing enzymes onto cellulose and its derived matrices. These methodologies can be extensively characterized as covalent and physical processes in accordance with the molecular forces among matrices and enzymes [21].

Few native cellulose materials can be utilized as a matrix for immobilization of enzymes. Loofah sponge has been investigated for covalent immobilization of lipase. Cellulose in loofah sponge was oxidized through sodium periodate to add aldehyde groups and then combined with lipase. The immobilized form of lipase showed enhanced thermal immovability or reprocessing. Periodate oxidation method has also been used to obtain cellulose derivatives. Periodate-oxidized cellulose acetate (an ideal substrate for enzyme immobilization) membrane has been used as a matrix for immobilization of cholesterol oxidases or enzyme binding membrane to synthesize a fiber-optic fluorescent biosensor [109].

Adjacent to above-mentioned substances of cellulose; nanoscale materials have been perfect for immobilization of enzymes. As it was complicated to improve nanoscale components, therefore, few changes were made utilizing magnetic cellulose nano (MCN) particles synthesized through adding magnetite. Covalent binding of α-amylase to MCN or magnetic cellulose nanoparticles ended in the creation of an innovative starch degrading system [110].

4.7 Chemical modification of covalent immobilization of enzyme onto cellulose nano fiber support

Mainly, the covalent attachment is a well-renowned process, depending on the chemical reaction of the side chain, the attachment between active amino groups on the surface of enzyme and active functionality that join onto cellulose nano fibers support surface. Before attachment preparation, the connection of covalent attachment might be attained through activation of surface groups on cellulose nano fibers support and enzyme molecule. Frequently, the activation of a surface functional group on cellulose fiber support gives very effective relation with enzyme throughout immobilization. In some cases, mode of attachment or different bonds on a reactive group of enzymes is essential. Furthermore, the surface variation of enzyme should be examined prior to the attachment of support that shows a complicated or invalidates covalent interface process. Recent investigations barely focus on the commencement of surface functional groups of cellulose nano fiber support and on chemical changes through utilizing chemical combining agent.

Covalent immobilization of enzyme onto cellulose nano fiber support belongs to one of the basic natures which do not just reveal the inactivation of enzyme active site but also probably irreversible attachment reaction, modification of the reactive group of enzymes for the reason of chemical changes and misdirection of enzyme orientation [111]. On the other hand, these drawbacks can be avoided by appropriate care in chemical modification or pretreatment. Enzyme denaturation can be through precise covalent immobilization by providing an oriented and reproducible control.

Cellulose nano fibers have great potential to be implemented in nano-biotechnology particularly in applications of immobilization of enzyme. Enzyme immobilization onto an innate-dependent support, specifically cellulose nano fiber has the capability for enhancing production yield and enzymatic performance, in addition to participating in green technology and sustainable sources. The cellulose nano fiber surface contains mostly – OH functional group that can be openly connected inadequately with enzyme and its attachment can be enhanced via surface alteration and interaction of chemical combination that creates a powerful and firm covalent immobilization of enzyme. The use of cellulose nano fiber in nano-biocatalyst has also the potential to be implemented in great applications, involving biosensors, medical diagnostics, pharmaceuticals, food and agriculture industries [112].

Conclusion

The process of immobilization has been used for improving the activity and stability of the enzyme in aqueous and non-aqueous media. In enzyme immobilization, selection and designing the support matrix is very important. For stabilization of enzyme, the use of nanoparticles has emerged as a versatile tool for generating excellent supports because of their small size and large surface area. In spite of current progresses in enzyme immobilization onto magnetic nanoparticles, these systems (i.e., covalent attachment, entrapment, and adsorption) are still suffering from alteration in properties, mass transfer restrictions, and low efficiency against insoluble substrates. Additional research is required to solve these problems for long-lasting industrial applications of immobilized enzymes.

References

[1] Y. Bai, H. Huang, K. Meng, P. Shi, P. Yang, H. Luo, C. Luo, Y. Feng, W. Zhang, B. Yao, Identification of an acidic α-amylase from Alicyclobacillus sp. A4 and assessment of its application in the starch industry, Food Chem. 131 (2012) 1473–1478. https://doi.org/10.1016/j.foodchem.2011.10.036

[2] B. Ismail, S.S. Nielsen, Invited review: plasmin protease in milk: current knowledge and relevance to dairy industry, J. Dairy Sci. 93 (2010) 4999–5009. https://doi.org/10.3168/jds.2010-3122

[3] R. DiCosimo, J. McAuliffe, A.J. Poulose, G. Bohlmann, Industrial use of immobilized enzymes, Chem. Soc. Rev. 42 (2013) 6437–6474. https://doi.org/10.1039/c3cs35506c

[4] C.R. Gomes-Ruffi, R.H. da Cunha, E.L. Almeida, Y.K. Chang, C.J. Steel, Effect of the emulsifier sodium stearoyl lactylate and of the enzyme maltogenic amylase on the quality of pan bread during storage, LWT-Food Sci. Technol. 49 (2012) 96–101. https://doi.org/10.1016/j.lwt.2012.04.014

[5] D. Jaros, H. Rohm, Enzymes Exogenous to Milk in Dairy Technology: Transglutaminase, (2015).

[6] J. Schückel, A. Matura, K.-H. Van Pee, One-copper laccase-related enzyme from Marasmius sp.: Purification, characterization and bleaching of textile dyes, Enzyme Microb. Technol. 48 (2011) 278–284. https://doi.org/10.1016/j.enzmictec.2010.12.002

[7] C.S. Rao, T. Sathish, P. Ravichandra, R.S. Prakasham, Characterization of thermo- and detergent stable serine protease from isolated Bacillus circulans and evaluation

of eco-friendly applications, Process Biochem. 44 (2009) 262–268.
https://doi.org/10.1016/j.procbio.2008.10.022

[8]   T.K. Hakala, T. Liitiä, A. Suurnäkki, Enzyme-aided alkaline extraction of
       oligosaccharides and polymeric xylan from hardwood kraft pulp, Carbohydr.
       Polym. 93 (2013) 102–108. https://doi.org/10.1016/j.carbpol.2012.05.013

[9]   I.M. Apetrei, M.L. Rodriguez-Mendez, C. Apetrei, J.A. De Saja, Enzyme sensor
       based on carbon nanotubes/cobalt (II) phthalocyanine and tyrosinase used in
       pharmaceutical analysis, Sensors Actuators B Chem. 177 (2013) 138–144.
       https://doi.org/10.1016/j.snb.2012.10.131

[10]   R. Das, S. Ghosh, C. Bhattacharjee, Enzyme membrane reactor in isolation of
       antioxidative peptides from oil industry waste: A comparison with non-peptidic
       antioxidants, LWT-Food Sci. Technol. 47 (2012) 238–245.
       https://doi.org/10.1016/j.lwt.2012.01.011

[11]   K. Luo, Q.I. Yang, J. Yu, X. Li, G. Yang, B. Xie, F. Yang, W. Zheng, G. Zeng,
       Combined effect of sodium dodecyl sulfate and enzyme on waste activated sludge
       hydrolysis and acidification, Bioresour. Technol. 102 (2011) 7103–7110.
       https://doi.org/10.1016/j.biortech.2011.04.023

[12]   S. Akhtar, Q. Husain, Potential applications of immobilized bitter gourd
       (Momordica charantia) peroxidase in the removal of phenols from polluted water,
       Chemosphere. 65 (2006) 1228–1235.
       https://doi.org/10.1016/j.chemosphere.2006.04.049

[13]   D. Tonini, T. Astrup, Life-cycle assessment of a waste refinery process for
       enzymatic treatment of municipal solid waste, Waste Manag. 32 (2012) 165–176.
       https://doi.org/10.1016/j.wasman.2011.07.027

[14]   I.M. Atadashi, M.K. Aroua, A.A. Aziz, High quality biodiesel and its diesel engine
       application: a review, Renew. Sustain. Energy Rev. 14 (2010) 1999–2008.
       https://doi.org/10.1016/j.rser.2010.03.020

[15]   A.D. Smith, S.P. Datta, G.H. Smith, Oxford dictionary of biochemistry and
       molecular biology, Oxford University Press, 1997.

[16]   A. Bairoch, The ENZYME database in 2000, Nucleic Acids Res. 28 (2000) 304–
       305. https://doi.org/10.1093/nar/28.1.304

[17]   K.M. Koeller, C.-H. Wong, Enzymes for chemical synthesis, Nature. 409 (2001)
       232. https://doi.org/10.1038/35051706

[18] C.-H. Wong, G.M. Whitesides, Enzymes in synthetic organic chemistry, Elsevier, 1994.

[19] C. Mateo, J.M. Palomo, G. Fernandez-Lorente, J.M. Guisan, R. Fernandez-Lafuente, Improvement of enzyme activity, stability and selectivity via immobilization techniques, Enzyme Microb. Technol. 40 (2007) 1451–1463. https://doi.org/10.1016/j.enzmictec.2007.01.018

[20] E. Katchalski-Katzir, D.M. Kraemer, Eupergit® C, a carrier for immobilization of enzymes of industrial potential, J. Mol. Catal. B Enzym. 10 (2000) 157–176. https://doi.org/10.1016/S1381-1177(00)00124-7

[21] R.A. Sheldon, Enzyme immobilization: the quest for optimum performance, Adv. Synth. Catal. 349 (2007) 1289–1307. https://doi.org/10.1002/adsc.200700082

[22] I. Eş, J.D.G. Vieira, A.C. Amaral, Principles, techniques, and applications of biocatalyst immobilization for industrial application, Appl. Microbiol. Biotechnol. 99 (2015) 2065–2082. https://doi.org/10.1007/s00253-015-6390-y

[23] J.E. Leresche, H.-P. Meyer, Chemocatalysis and biocatalysis (biotransformation): some thoughts of a chemist and of a biotechnologist, Org. Process Res. Dev. 10 (2006) 572–580. https://doi.org/10.1021/op0600308

[24] W. Hartmeier, Immobilized biocatalysts—from simple to complex systems, Trends Biotechnol. 3 (1985) 149–153. https://doi.org/10.1016/0167-7799(85)90104-0

[25] P. V Iyer, L. Ananthanarayan, Enzyme stability and stabilization—aqueous and non-aqueous environment, Process Biochem. 43 (2008) 1019–1032. https://doi.org/10.1016/j.procbio.2008.06.004

[26] S. Spisak, A. Guttman, Biomedical applications of protein microarrays, Curr. Med. Chem. 16 (2009) 2806–2815. https://doi.org/10.2174/092986709788803141

[27] D.S. Clark, Can immobilization be exploited to modify enzyme activity?, Trends Biotechnol. 12 (1994) 439–443. https://doi.org/10.1016/0167-7799(94)90018-3

[28] S. Shanmugam, Enzyme technology, IK International Pvt Ltd, 2009.

[29] G.T.R. Kulkarni, Biotechnology and its applications in pharmacy, Jaypee Bros Medical Publishers, 2002.

[30] P.K. Jena, M. Rath, C.S. Mishra, K. Melal and Mineral Recovery Through Bioleaching, Biotechnol. Appl. Mishra, CSK, Champagne, P., Eds (IK Int. Publ. House Pvt. Ltd., New Delhi, India). (2009) 309–319.

Materials Research Forum LLC
doi: http://dx.doi.org/10.21741/9781644900079-1

[31]   P.N. Krishna, Enzyme technology: pacemaker of biotechnology, PHI Learning Pvt. Ltd., 2011.

[32]   J.M. Nelson, E.G. Griffin, Adsorption of Invertase, J. Am. Chem. Soc. 38 (1916) 1109–1115. https://doi.org/10.1021/ja02262a018

[33]   D. Brady, J. Jordaan, Advances in enzyme immobilisation, Biotechnol. Lett. 31 (2009) 1639-1650. https://doi.org/10.1007/s10529-009-0076-4

[34]   S. Cantone, V. Ferrario, L. Corici, C. Ebert, D. Fattor, P. Spizzo, L. Gardossi, Efficient immobilisation of industrial biocatalysts: criteria and constraints for the selection of organic polymeric carriers and immobilisation methods, Chem. Soc. Rev. 42 (2013) 6262–6276. https://doi.org/10.1039/c3cs35464d

[35]   J.R. Cherry, A.L. Fidantsef, Directed evolution of industrial enzymes: an update, Curr. Opin. Biotechnol. 14 (2003) 438–443. https://doi.org/10.1016/S0958-1669(03)00099-5

[36]   G. Massolini, E. Calleri, Immobilized trypsin systems coupled online to separation methods: Recent developments and analytical applications, J. Sep. Sci. 28 (2005) 7–21. https://doi.org/10.1002/jssc.200401941

[37]   U. Guzik, K. Hupert-Kocurek, D. Wojcieszyńska, Immobilization as a strategy for improving enzyme properties-application to oxidoreductases, Molecules. 19 (2014) 8995–9018. https://doi.org/10.3390/molecules19078995

[38]   R.C. Rodrigues, C. Ortiz, Á. Berenguer-Murcia, R. Torres, R. Fernández-Lafuente, Modifying enzyme activity and selectivity by immobilization, Chem. Soc. Rev. 42 (2013) 6290–6307. https://doi.org/10.1039/C2CS35231A

[39]   F. Secundo, Conformational changes of enzymes upon immobilisation, Chem. Soc. Rev. 42 (2013) 6250–6261. https://doi.org/10.1039/c3cs35495d

[40]   A. Liese, L. Hilterhaus, Evaluation of immobilized enzymes for industrial applications, Chem. Soc. Rev. 42 (2013) 6236–6249. https://doi.org/10.1039/c3cs35511j

[41]   A. Ursini, P. Maragni, C. Bismara, B. Tamburini, Enzymatic method of preparation of opticallly active trans-2-amtno cyclohexanol derivatives, Synth. Commun. 29 (1999) 1369–1377. https://doi.org/10.1080/00397919908086112

[42]   M. Goto, C. Hatanaka, M. Goto, Immobilization of surfactant–lipase complexes and their high heat resistance in organic media, Biochem. Eng. J. 24 (2005) 91–94. https://doi.org/10.1016/j.bej.2005.01.027

[43] M.H.A. Janssen, L.M. van Langen, S.R.M. Pereira, F. van Rantwijk, R.A. Sheldon, Evaluation of the performance of immobilized penicillin G acylase using active site titration, Biotechnol. Bioeng. 78 (2002) 425–432. https://doi.org/10.1002/bit.10208

[44] R.C. Rodrigues, A. Berenguer-Murcia, R. Fernandez-Lafuente, Coupling chemical modification and immobilization to improve the catalytic performance of enzymes, Adv. Synth. Catal. 353 (2011) 2216–2238. https://doi.org/10.1002/adsc.201100163

[45] R.A. Sheldon, S. van Pelt, Enzyme immobilisation in biocatalysis: why, what and how, Chem. Soc. Rev. 42 (2013) 6223–6235. https://doi.org/10.1039/C3CS60075K

[46] R. Ahmad, M. Sardar, Enzyme immobilization: an overview on nanoparticles as immobilization matrix, Biochem. Anal. Biochem. 4 (2015) 1.

[47] C. Spahn, S.D. Minteer, Enzyme immobilization in biotechnology, Recent Patents Eng. 2 (2008) 195–200. https://doi.org/10.2174/187221208786306333

[48] R.D. Johnson, Z.-G. Wang, F.H. Arnold, Surface site heterogeneity and lateral interactions in multipoint protein adsorption, J. Phys. Chem. 100 (1996) 5134–5139. https://doi.org/10.1021/jp9523682

[49] S.F. D'souza, Immobilized enzymes in bioprocess, Curr. Sci. (1999) 69–79.

[50] J. Fu, J. Reinhold, N.W. Woodbury, Peptide-modified surfaces for enzyme immobilization, PLoS One. 6 (2011) e18692. https://doi.org/10.1371/journal.pone.0018692

[51] V. Singh, M. Sardar, M.N. Gupta, Immobilization of enzymes by bioaffinity layering, in: Immobil. Enzym. Cells, Springer, 2013: pp. 129–137. https://doi.org/10.1007/978-1-62703-550-7_9

[52] M. Hartmann, X. Kostrov, Immobilization of enzymes on porous silicas–benefits and challenges, Chem. Soc. Rev. 42 (2013) 6277–6289. https://doi.org/10.1039/c3cs60021a

[53] Z. Grosová, M. Rosenberg, M. Rebroš, M. Šipocz, B. Sedláčková, Entrapment of β-galactosidase in polyvinylalcohol hydrogel, Biotechnol. Lett. 30 (2008) 763–767. https://doi.org/10.1007/s10529-007-9606-0

[54] A. Deshpande, S.F. D'souza, G.B. Nadkarni, Coimmobilization of D-amino acid oxidase and catalase by entrapment ofTrigonopsis variabilis in radiation

polymerised Polyacrylamide beads, J. Biosci. 11 (1987) 137–144.
https://doi.org/10.1007/BF02704664

[55] I. Galaev, B. Mattiasson, Smart polymers for bioseparation and bioprocessing,
CRC Press, 2001.

[56] H.H.P. Yiu, M.A. Keane, Enzyme–magnetic nanoparticle hybrids: new effective
catalysts for the production of high value chemicals, J. Chem. Technol.
Biotechnol. 87 (2012) 583–594. https://doi.org/10.1002/jctb.3735

[57] Y. Zhou, S. Pan, X. Wei, L. Wang, Y. Liu, Immobilization of β-glucosidase onto
magnetic nanoparticles and evaluation of the enzymatic properties, BioResources.
8 (2013) 2605–2619. https://doi.org/10.15376/biores.8.2.2605-2619

[58] J. Xu, J. Sun, Y. Wang, J. Sheng, F. Wang, M. Sun, Application of iron magnetic
nanoparticles in protein immobilization, Molecules. 19 (2014) 11465–11486.
https://doi.org/10.3390/molecules190811465

[59] S.A. Ansari, Q. Husain, Potential applications of enzymes immobilized on/in nano
materials: a review, Biotechnol. Adv. 30 (2012) 512–523.
https://doi.org/10.1016/j.biotechadv.2011.09.005

[60] R. Singh, M. Tiwari, R. Singh, J.-K. Lee, From protein engineering to
immobilization: promising strategies for the upgrade of industrial enzymes, Int. J.
Mol. Sci. 14 (2013) 1232–1277. https://doi.org/10.3390/ijms14011232

[61] M. Soleimani, A. Khani, K. Najafzadeh, α-Amylase immobilization on the silica
nanoparticles for cleaning performance towards starch soils in laundry detergents,
J. Mol. Catal. B Enzym. 74 (2012) 1–5.
https://doi.org/10.1016/j.molcatb.2011.07.011

[62] M.L. Verma, R. Chaudhary, T. Tsuzuki, C.J. Barrow, M. Puri, Immobilization of
β-glucosidase on a magnetic nanoparticle improves thermostability: application in
cellobiose hydrolysis, Bioresour. Technol. 135 (2013) 2–6.
https://doi.org/10.1016/j.biortech.2013.01.047

[63] K. Thandavan, S. Gandhi, S. Sethuraman, J.B.B. Rayappan, U.M. Krishnan,
Development of electrochemical biosensor with nano-interface for xanthine
sensing–A novel approach for fish freshness estimation, Food Chem. 139 (2013)
963–969. https://doi.org/10.1016/j.foodchem.2013.02.008

[64] L. Wei, W. Zhang, H. Lu, P. Yang, Immobilization of enzyme on detonation
nanodiamond for highly efficient proteolysis, Talanta. 80 (2010) 1298–1304.
https://doi.org/10.1016/j.talanta.2009.09.029

Materials Research Forum LLC
doi: http://dx.doi.org/10.21741/9781644900079-1

[65]   Y.Z. Chen, C.T. Yang, C.B. Ching, R. Xu, Immobilization of lipases on hydrophobilized zirconia nanoparticles: highly enantioselective and reusable biocatalysts, Langmuir. 24 (2008) 8877–8884. https://doi.org/10.1021/la801384c

[66]   X. Meng, G. Xu, Q.-L. Zhou, J.-P. Wu, L.-R. Yang, Highly efficient solvent-free synthesis of 1, 3-diacylglycerols by lipase immobilised on nano-sized magnetite particles, Food Chem. 143 (2014) 319–324. https://doi.org/10.1016/j.foodchem.2013.07.132

[67]   H.-H. Yang, S.-Q. Zhang, X.-L. Chen, Z.-X. Zhuang, J.-G. Xu, X.-R. Wang, Magnetite-containing spherical silica nanoparticles for biocatalysis and bioseparations, Anal. Chem. 76 (2004) 1316–1321. https://doi.org/10.1021/ac034920m

[68]   G.K. Kouassi, J. Irudayaraj, G. McCarty, Examination of cholesterol oxidase attachment to magnetic nanoparticles, J. Nanobiotechnology. 3 (2005) 1. https://doi.org/10.1186/1477-3155-3-1

[69]   G.A. Petkova, K. Záruba, P. Žvátora, V. Král, Gold and silver nanoparticles for biomolecule immobilization and enzymatic catalysis, Nanoscale Res. Lett. 7 (2012) 287.

[70]   A.K. Johnson, A.M. Zawadzka, L.A. Deobald, R.L. Crawford, A.J. Paszczynski, Novel method for immobilization of enzymes to magnetic nanoparticles, J. Nanoparticle Res. 10 (2008) 1009–1025. https://doi.org/10.1007/s11051-007-9332-5

[71]   N. Miletić, V. Abetz, K. Ebert, K. Loos, Immobilization of Candida antarctica lipase B on polystyrene nanoparticles, Macromol. Rapid Commun. 31 (2010) 71–74. https://doi.org/10.1002/marc.200900497

[72]   A.Y. Chen, Z. Deng, A.N. Billings, U.O.S. Seker, M.Y. Lu, R.J. Citorik, B. Zakeri, T.K. Lu, Synthesis and patterning of tunable multiscale materials with engineered cells, Nat. Mater. 13 (2014) 515. https://doi.org/10.1038/nmat3912

[73]   P.Q. Nguyen, Z. Botyanszki, P.K.R. Tay, N.S. Joshi, Programmable biofilm-based materials from engineered curli nanofibres, Nat. Commun. 5 (2014) 4945. https://doi.org/10.1038/ncomms5945

[74]   J.W. Costerton, Z. Lewandowski, D.E. Caldwell, D.R. Korber, H.M. Lappin-Scott, Microbial biofilms, Annu. Rev. Microbiol. 49 (1995) 711–745. https://doi.org/10.1146/annurev.mi.49.100195.003431

Materials Research Foundations **44** (2019) 1-28

doi: http://dx.doi.org/10.21741/9781644900079-1

[75]   H.H.P. Fang, L.-C. Xu, K.-Y. Chan, Effects of toxic metals and chemicals on biofilm and biocorrosion, Water Res. 36 (2002) 4709–4716. https://doi.org/10.1016/S0043-1354(02)00207-5

[76]   R. Gross, K. Lang, K. Bühler, A. Schmid, Characterization of a biofilm membrane reactor and its prospects for fine chemical synthesis, Biotechnol. Bioeng. 105 (2010) 705–717.

[77]   V. Sivanathan, A. Hochschild, Generating extracellular amyloid aggregates using E. coli cells, Genes Dev. (2012) *2659-2667.* https://doi.org/10.1101/gad.205310.112

[78]   V. Sivanathan, A. Hochschild, A bacterial export system for generating extracellular amyloid aggregates, Nat. Protoc. 8 (2013) 1381-1391. https://doi.org/10.1038/nprot.2013.081

[79]   B. Zakeri, J.O. Fierer, E. Celik, E.C. Chittock, U. Schwarz-Linek, V.T. Moy, M. Howarth, Peptide tag forming a rapid covalent bond to a protein, through engineering a bacterial adhesin, Proc. Natl. Acad. Sci. 109 (2012) E690–E697. https://doi.org/10.1073/pnas.1115485109

[80]   Z. Botyanszki, P.K.R. Tay, P.Q. Nguyen, M.G. Nussbaumer, N.S. Joshi, Engineered catalytic biofilms: Site-specific enzyme immobilization onto E. coli curli nanofibers, Biotechnol. Bioeng. 112 (2015) 2016–2024. https://doi.org/10.1002/bit.25638

[81]   C. Altinkaynak, S. Tavlasoglu, I. Ocsoy, A new generation approach in enzyme immobilization. Organic-inorganic hybrid nanoflowers with enhanced catalytic activity and stability, Enzyme Microb. Technol. 93 (2016) 105–112. https://doi.org/10.1016/j.enzmictec.2016.06.011

[82]   C. Ji, L.N. Nguyen, J. Hou, F.I. Hai, V. Chen, Direct immobilization of laccase on titania nanoparticles from crude enzyme extracts of P. ostreatus culture for micro-pollutant degradation, Sep. Purif. Technol. 178 (2017) 215–223. https://doi.org/10.1016/j.seppur.2017.01.043

[83]   L. Lloret, G. Eibes, G. Feijoo, M.T. Moreira, J.M. Lema, F. Hollmann, Immobilization of laccase by encapsulation in a sol–gel matrix and its characterization and use for the removal of estrogens, Biotechnol. Prog. 27 (2011) 1570–1579. https://doi.org/10.1002/btpr.694

[84]   J. Hou, G. Dong, B. Luu, R.G. Sengpiel, Y. Ye, M. Wessling, V. Chen, Hybrid membrane with TiO$_2$ based bio-catalytic nanoparticle suspension system for the

degradation of bisphenol-A, Bioresour. Technol. 169 (2014) 475–483. https://doi.org/10.1016/j.biortech.2014.07.031

[85] C. Ji, J. Hou, K. Wang, Y. Zhang, V. Chen, Biocatalytic degradation of carbamazepine with immobilized laccase-mediator membrane hybrid reactor, J. Memb. Sci. 502 (2016) 11–20. https://doi.org/10.1016/j.memsci.2015.12.043

[86] R. Ricco, C. Pfeiffer, K. Sumida, C.J. Sumby, P. Falcaro, S. Furukawa, N.R. Champness, C.J. Doonan, Emerging applications of metal–organic frameworks, CrystEngComm. 18 (2016) 6532–6542. https://doi.org/10.1039/C6CE01030J

[87] X. Wu, M. Hou, J. Ge, Metal–organic frameworks and inorganic nanoflowers: a type of emerging inorganic crystal nanocarrier for enzyme immobilization, Catal. Sci. Technol. 5 (2015) 5077–5085. https://doi.org/10.1039/C5CY01181G

[88] R.W. Larsen, L. Wojtas, J. Perman, R.L. Musselman, M.J. Zaworotko, C.M. Vetromile, Mimicking heme enzymes in the solid state: metal–organic materials with selectively encapsulated heme, J. Am. Chem. Soc. 133 (2011) 10356–10359. https://doi.org/10.1021/ja203068u

[89] Y. Chen, V. Lykourinou, C. Vetromile, T. Hoang, L.-J. Ming, R.W. Larsen, S. Ma, How can proteins enter the interior of a MOF? Investigation of cytochrome c translocation into a MOF consisting of mesoporous cages with microporous windows, J. Am. Chem. Soc. 134 (2012) 13188–13191. https://doi.org/10.1021/ja305144x

[90] X. Lian, Y.-P. Chen, T.-F. Liu, H.-C. Zhou, Coupling two enzymes into a tandem nanoreactor utilizing a hierarchically structured MOF, Chem. Sci. 7 (2016) 6969–6973. https://doi.org/10.1039/C6SC01438K

[91] W. Liu, C. Wu, C. Chen, B. Singco, C. Lin, H. Huang, Fast multipoint immobilized MOF bioreactor, Chem. Eur. J. 20 (2014) 8923–8928. https://doi.org/10.1002/chem.201400270

[92] W. Liu, N. Yang, Y. Chen, S. Lirio, C. Wu, C. Lin, H. Huang, Lipase-supported metal–organic framework bioreactor catalyzes warfarin synthesis, Chem. Eur. J. 21 (2015) 115–119. https://doi.org/10.1002/chem.201405252

[93] F.-K. Shieh, S.-C. Wang, C.-I. Yen, C.-C. Wu, S. Dutta, L.-Y. Chou, J. V Morabito, P. Hu, M.-H. Hsu, K.C.-W. Wu, Imparting functionality to biocatalysts via embedding enzymes into nanoporous materials by a de novo approach: size-selective sheltering of catalase in metal–organic framework microcrystals, J. Am. Chem. Soc. 137 (2015) 4276–4279. https://doi.org/10.1021/ja513058h

[94]   K. Liang, R. Ricco, C.M. Doherty, M.J. Styles, S. Bell, N. Kirby, S. Mudie, D. Haylock, A.J. Hill, C.J. Doonan, Biomimetic mineralization of metal-organic frameworks as protective coatings for biomacromolecules, Nat. Commun. 6 (2015) 7240. https://doi.org/10.1038/ncomms8240

[95]   M. Sánchez-Sánchez, N. Getachew, K. Díaz, M. Díaz-García, Y. Chebude, I. Díaz, Synthesis of metal–organic frameworks in water at room temperature: salts as linker sources, Green Chem. 17 (2015) 1500–1509. https://doi.org/10.1039/C4GC01861C

[96]   D. Ruano, M. Díaz-García, A. Alfayate, M. Sánchez-Sánchez, Nanocrystalline M-MOF-74 as heterogeneous catalysts in the oxidation of cyclohexene: Correlation of the activity and redox potential, ChemCatChem. 7 (2015) 674–681. https://doi.org/10.1002/cctc.201402927

[97]   X. Wu, J. Ge, C. Yang, M. Hou, Z. Liu, Facile synthesis of multiple enzyme-containing metal-organic frameworks in a biomolecule-friendly environment, Chem. Commun. 51 (2015) 13408–13411. https://doi.org/10.1039/C5CC05136C

[98]   Y. Chen, S. Ma, Biomimetic catalysis of metal-organic frameworks, Dalt. Trans. 45 (2016) 9744–9753. https://doi.org/10.1039/C6DT00325G

[99]   P.A. Sontz, J.B. Bailey, S. Ahn, F.A. Tezcan, A metal organic framework with spherical protein nodes: Rational chemical design of 3D protein crystals, J. Am. Chem. Soc. 137 (2015) 11598–11601. https://doi.org/10.1021/jacs.5b07463

[100] D. Fujita, M. Fujita, Fitting proteins into metal organic frameworks, ACS Cent. Sci. 1 (2015) 352–353. https://doi.org/10.1021/acscentsci.5b00315

[101] X. Wu, C. Yang, J. Ge, Z. Liu, Polydopamine tethered enzyme/metal-organic framework composites with high stability and reusability, Nanoscale. 7 (2015) 18883–18886. https://doi.org/10.1039/C5NR05190H

[102] J. Shi, X. Wang, S. Zhang, L. Tang, Z. Jiang, Enzyme-conjugated ZIF-8 particles as efficient and stable Pickering interfacial biocatalysts for biphasic biocatalysis, J. Mater. Chem. B. 4 (2016) 2654–2661. https://doi.org/10.1039/C6TB00104A

[103] V. Gascón, E. Castro-Miguel, M. Díaz-García, R.M. Blanco, M. Sanchez-Sanchez, In situ and post-synthesis immobilization of enzymes on nanocrystalline MOF platforms to yield active biocatalysts, J. Chem. Technol. Biotechnol. 92 (2017) 2583–2593. https://doi.org/10.1002/jctb.5274

[104] D. Klemm, B. Heublein, H.P. Fink, A. Bohn, Cellulose: Fascinating biopolymer and sustainable raw material, Angew. Chemie - Int. Ed. 44 (2005) 3358–3393. https://doi.org/10.1002/anie.200460587

[105] I. Siró, D. Plackett, Microfibrillated cellulose and new nanocomposite materials: A review, Cellulose. 17 (2010) 459–494. https://doi.org/10.1007/s10570-010-9405-y

[106] S. Iwamoto, A.N. Nakagaito, H. Yano, Nano-fibrillation of pulp fibers for the processing of transparent nanocomposites, Appl. Phys. A Mater. Sci. Process. 89 (2007) 461–466. https://doi.org/10.1007/s00339-007-4175-6

[107] S. Zhu, Y. Wu, Q. Chen, Z. Yu, C. Wang, S. Jin, Y. Ding, G. Wu, Dissolution of cellulose with ionic liquids and its application: A mini-review, Green Chem. 8 (2006) 325–327. https://doi.org/10.1039/b601395c

[108] X. Wu, F. Zhao, J.R. Varcoe, A.E. Thumser, C. Avignone-Rossa, R.C.T. Slade, Direct electron transfer of glucose oxidase immobilized in an ionic liquid reconstituted cellulose-carbon nanotube matrix, Bioelectrochemistry. 77 (2009) 64–68. https://doi.org/10.1016/j.bioelechem.2009.05.008

[109] S. Wang, S. Li, Y. Yu, Immobilization of cholesterol oxidase on cellulose acetate membrane for free cholesterol biosensor development, Artif. Cells. Blood Substit. Immobil. Biotechnol. 32 (2004) 413–425. https://doi.org/10.1081/BIO-200027479

[110] M. Namdeo, S.K. Bajpai, Immobilization of α-amylase onto cellulose-coated magnetite (CCM) nanoparticles and preliminary starch degradation study, J. Mol. Catal. B Enzym. 59 (2009) 134–139. https://doi.org/10.1016/j.molcatb.2009.02.005

[111] F. Rusmini, Z. Zhong, J. Feijen, Protein immobilization strategies for protein biochips, Biomacromolecules. 8 (2007) 1775–1789. https://doi.org/10.1021/bm061197b

[112] S. Sulaiman, M.N. Mokhtar, M.N. Naim, A.S. Baharuddin, A. Sulaiman, A Review: Potential Usage of cellulose nanofibers (CNF) for enzyme immobilization via covalent interactions, Appl. Biochem. Biotechnol. 175 (2014) 1817–1842. https://doi.org/10.1007/s12010-014-1417-x

Enzymatic Fuel Cells
Materials Research Foundations **44** (2019) 28-50

Materials Research Forum LLC
doi: http://dx.doi.org/10.21741/9781644900079-2

Chapter 2

Use of Enzymes in Different Types of Biofuel Cells

Amna Ahmad[1], Muhammad Hussnain Siddique[1], Saima Muzammil[2], Muhammad Riaz[3], Muhammad Waseem[2], Ijaz Rasul[1], Farrukh Azeem[1], Sabir Hussain[4], Habibullah Nadeem*[1]

[1]Department of Bioinformatics and Biotechnology, Government College University, Faisalabad, Pakistan

[2]Department of Microbiology, Government College University, Faisalabad, Pakistan

[3]Department of Food Sciences, University College of Agriculture, Bahauddin Zakariya University, Multan, Pakistan

[4]Department of Environmental Sciences and Engineering, Government College University, Faisalabad, Pakistan

* habibullah@gcuf.edu.pk

Abstract

Enzymes are those macromolecules which are needed for many chemical interconversion that maintain life. Progress in protein engineering and recombinant technology has developed the enzyme as essential molecules for use in therapeutical and industrial processes. For the conversion of chemical energy into electrical energy many electrochemical systems i.e., biofuel cells have used enzyme as their catalysts. In this chapter, we focused on the different type of biofuel cells which use enzymes for energy conversion. Enzymes, microbes and organelles are the biocatalysts which can be employed in biofuel cells.

There are vast applications of biofuel cells. These devices might be employed as embedable power sources, where the cells get power via a fuel like glucose in the bloodstream because of their biocompatible catalysts. To make biofuel cells a viable device for unconventional production of energy there are some disadvantages of biofuel cells which should be taken into account. Many enzymes in biofuel cells need mediators or specific electrode surface to proficiently permit the flow of electron from enzymes to electrodes because electron flow to the electrode from enzymes is intrinsically aberrant. A further disadvantage of biofuel cells is that enzymes are quite unsteady catalysts in the long interval and their increasing deprivation in constant functioning end up in continuing power loss. Conversely major steps in the stabilization of enzymes have been made;

Enzymatic Fuel Cells
Materials Research Foundations 44 (2019) 28-50

Materials Research Forum LLC
doi: http://dx.doi.org/10.21741/9781644900079-2

possible and efficient solutions to this difficulty contain genetic modification, enzyme regeneration in situ [17] and micellar enzyme encapsulation [18].

In a range of feasible insert able clinical gadgets the fuel and oxygen needed for their function can feasibly be taken from their instant environment which recommends high potential as a power source, therefore these cells can be inserted in a living system.

Keywords

Glucose Oxidase, Bilirubin Oxidase, Bioelectrochemical System

Contents

Materials Research Forum LLC
doi: http://dx.doi.org/10.21741/9781644900079-2

1.   History of Enzymes

A physiology professor named as Wilhelm Friedrich Kuhne at Heidelberg University used the word enzyme in 1877 which was derived from a Greek word "in leaven". However, the term leaven is used to refer a substance that causes fermentation in dough [1]. Though people were well aware with enzymes and their uses since centuries, Wilhelm Friedrich Kuhne was the scientist who conferred a scientific terminology to the biological macro-molecules [2]. Early, Egyptians have used enzymes for the preservation purpose of beverages and food. Until 400 BC, cheese manufacturing has use enzymes for eternity, when Homer's Iliad revealed the use of stomach of the kid for the synthesis of cheese. The significance of this macro-molecule was described by a famous Italian catholic priest Lazzaro Spallanzani in his work of biogenesis meaning impulsive production of microbes in 1738. In his work, he revealed that there was a force for the generation of life innate to several types of inorganic materials which make living microbes to generate themselves inadequate duration [3]. The process of changing starch into glucose was examined by Gottlieb Sigismund Kirchhof in 1812. He also explained the applications of these macro-molecules as catalysts in his investigation. Anselem Payen was the first French chemist that discovered first enzyme diastase in 1833. Another Swedish scientist Jöns Jacob Berzelius recognized the hydrolysis of starch in the presence of diastase as a catalytic reaction in 1835. He also inferred the fermentation process in 1839 occurred by a catalytic force and assumed that a body by its simple existence, through resemblance to the fermentable substance, made its reorganization into the product [4]. A process involving the versatile attachment of substrate on enzyme at an elevated concentration of sucrose was applied for anticompetitive inhibition, elucidated via invertase derived from *S. cervisiae* [5].

2.   Introduction

Enzymes are large bio-molecules or biological substances which are synthesized by living organisms and behave like a catalyst to carry out a particular biochemical reaction. Enzymes also act as chemical catalysts in a chemical reaction that aid to speed up the reactions within the cell or outside the cell. Generally, enzymes are also called as biocatalysts.

In early times, naturally available enzymes have been extensively used in the synthesis of products like leather, indigo and linen. All of these procedures based on the use of enzymes that exist in papaya fruit or generated through microorganisms. On large scale, fermentation procedures were improved specifically for the generation of well distinguished and purified enzymes by using specific strains. This progress facilitated the introduction of enzymes into real products and industrial processes used in textile, starch and detergent industries. The recombinant DNA technologies have endorsed the development of manufacturing procedures and provide help to generate those enzymes which could not be generated earlier on a commercial scale. Moreover, the progress in biotechnology like direct evolution and protein engineering has increased the advancement in the commercialization of industrially important enzymes.

Almost four thousand chemical reactions are catalyzed by the enzymes [6]. According to their nature enzymes are very precise and in 1894 it was recommended by Nobel laureate Emil Fischer that enzyme and substrate have particular complementary geometric shapes due to which they accurately fit into one another. This theory is generally called as lock and key model. Although, this model elaborated the specificity of enzymes but failed to describe the stabilization of transition state attained by enzymes.

Most of the industrial enzymes are hydrolytic in action therefore, they are used in the degradation of several natural substances. Because of their extreme use in the dairy and detergent industries proteases remain the prominent type of enzyme. Numerous carbohydrates primarily cellulases and amylases signify a major group of enzymes that are used in textile, starch, baking and detergent industries. Many functions are performed by enzymes in many processes and products that are generally come across in the synthesis of food and beverages, clothing, cleaning supplies, transportation fuels, paper products, monitoring devices and pharmaceuticals. Currently, hydrolases are the most commonly implemented enzymes in biotechnology which catalyze the breakdown of substances in the presence of water. For asymmetric manufacturing and racemic resolution, regional stereospecific properties present in enzymes have been exploited [2].

Enzymes are also used in many electrochemical systems like biofuel cells. Biofuel cells are that kind of energy conversion devices which applied biocatalysts like enzymes or entire living cells for the conversion of chemical energy into electrical energy. The normal fuels engaged in biofuel cells have been methanol, glucose, lactate and ethanol and an extensive range of fuels might be implemented because living cells have metabolic pathways for oxidizing an extensive range of substrates which include fatty acids, carbohydrates and alcohols. The classification of these cells depends on the type of biocatalysts used. Biocatalysts which can be employed in biofuel cells are of three kinds i.e., enzymes, microbes and organelles [7].

3.   Biofuel cells

A device that constantly produces electrical energy from accumulated chemical energy in a fuel cell at continuous fuel supply is known as a conventional fuel cell. Such fuel cells contain oxidant, fuel, cathodic and anodic substrate equipment. A semi-permeable membrane is usually used to separate the anodic and cathodic substrates. A biofuel cell consists of two platinum electrodes in the occurrence of catalyst where an apparent dissimilarity was observed among the two platinum electrodes as illustrated by Michael Cresse Potter in 1911 [8].

In 1912 Potter intimated the first biofuel cell containing yeast at cathode for the oxidation of glucose. Intimation of microbial fuel cells produces primary interest in biofuel cells, till the work of Kimble in 1960 that people planned to employ isolated enzymes at the surface of electrodes for the conversion of energy. Three diverse enzymatic bio-anodes e.g., one with an amino acid oxidase, second with alcohol dehydrogenase and third with glucose oxidase have been constructed and evaluated in three diverse biofuel cells made up of these bio-anodes created by Kimble and his co-workers in 1964. They have the potential to produce a positive open circuit for the two oxidases in complete biofuel cells and therefore, such work initiates the research area that we called enzymatic biofuel cells [9].

Researchers have been concurrently improving the biofuel in the last forty years. Those types of biofuel cells have been created which acquire duration more than five years. However, most of the microbial fuel cells have the potential to completely oxidize their fuel which is normally glucose or lactate although, few microbial fuel cells use wastewater and animal waste as their fuel, whereas some use carbohydrates obtained from the sandy soils from rivers and oceans. However, all of the biofuel cells have been restricted by fewer power densities and current [7, 10].

More protected than methanol fuel cells and Li-ion batteries, enzyme-based and microorganism based biofuel cells are producing bio-electricity by the oxidation of renewable sources like sugars and organic acids, in combination with the change of oxygen to water [11]. These cells become the next generation energy devices because of the fact that these bio-fuel devices are predicted to generate greater energy and permit an extensive variety of applications. Few problems arise if we consider the immobilization of the biocatalyst on electrode even if microorganisms and enzymes are highly efficient

Materials Research Forum LLC
doi: http://dx.doi.org/10.21741/9781644900079-2

biocatalysts. That's why microorganisms or solution borne enzymes are employed instead of immobilized biocatalyst on the surface of the electrode in microbial or enzymatic biofuel cells [8].

Taking as a substitute of conservative fuel cells, biofuel cells have many benefits which made them smart over conventional fuel cells. Biofuel cells do not normally experience poisoning and fuel crossover because they make use of enzymes as catalysts. These are the features which allow some of the biofuel to act in a mode with no section, and let effortless manufacture and avoid the necessity of a barrier membrane. Bio-fuel cells function in lenient temperature and pH states (25 to 37 °C and 5 to 8 pH) which are in bleak difference to the extremely basic or acidic and high temperature working condition of conventional bio-fuel cells. The fuels highlighted beyond that are normally employed in these cells are viable, more protective and simple to operate instead of conventional combustible materials like methanol, borohydride and hydrogen gas [12].

There are vast applications of biofuel cells. These devices might be employed as embedable power sources, where the cells get power via a fuel like glucose in the bloodstream because of their biocompatible catalysts. Based on wired glucose oxidase half of this objective has been partly recognized along through the implementation of subcutaneous (inserted inside the skin) sensor of glucose on wired glucose oxidase, however, these sensors should be substituted each 3 to 5 days which is not a satisfactory life for the substitution of a battery [13]. Just like insulin pumps and battery-less pacemakers, abovementioned applications have many exciting possibilities. Many obstacles such as biocompatibility and long-term stability of enzymes should be solved to attain implantable power. However, enzymes are eco-friendly, the immune system of the body would give a response to an inserted bio-fuel cell and those responses have to be contradicted or conscientious for the design of the cell. As verified by several statements, these cells functioning in plants [14] and animal [15], many developments have been on the function of biofuel cells inside living tissues, however, a device for durable human insertion is still far away [16].

To make biofuel cells a viable device for unconventional production of energy there are some disadvantages of biofuel cells which should be taken into account. Many enzymes in biofuel cells need mediators or specific electrode surface to proficiently permit the flow of electron from enzymes to electrodes because electron flow to the electrode from enzymes is intrinsically aberrant. A further disadvantage of biofuel cells is that enzymes are quite unsteady catalysts in the long interval and their increasing deprivation in constant functioning end up in continuing power loss. Conversely major steps in the stabilization of enzymes have been made; possible and efficient solutions to this difficulty contain genetic modification, enzyme regeneration in situ [17] and micellar enzyme encapsulation [18].

In a range of feasible insert able clinical gadgets the fuel and oxygen needed for their function can feasibly be taken from their instant environment which recommends high potential as a power source, therefore these cells can be inserted in the living system. For instance, in order to produce an electric current, a biosensor for glucose has been created employing a (glucose oxidase-dependent) anode and (cytochrome c) cathode [19]. Fuel cells have been exposed to produce energy through the oxidation of compounds obtained in wastewater rivulets. Two possible beneficial objectives can be achieved through this process; first is the elimination of organic compounds from the waste streams and second is the production of electric energy [20].

4.   Mechanism of biofuel cells

Biofuel cells depend on the enzyme catalysis for a minimum portion of their activity [21]. Furthermore, these fuel cells are able to directly changing the chemical energy into electrical energy through electrochemical reactions including biochemical processes.

Two reactions occur at the electrodes of the fuel cells are reduction reaction takes place at the cathode and an oxidation reaction takes place at the anode. Electrons are released by the oxidation which transfers to the cathode through the external circuit performing their electrical work. A rectify charge is moved as a result of the electrolyte frequently in the form of positive ions for the completion of the circuit.

Traditionally, these cells perform their function by using fuel like methanol or hydrogen and generate energy, carbon dioxide and water. Further fuels, like some minor sort of alkanes and alcohols, were also exploited and often transformed to generate hydrogen before the cells processes [22, 23].

In mild conditions (20 to 40 °C temperature or near neutral pH) biofuel cells utilized enzymes as catalysts (unaccompanied or in an organism). These features enable the cells to be used in conditions like at extreme heat, or where harsh circumstances in the reaction are objectionable. Additionally, different types of reactions capable to be catalyzed through enzymes formulate the exploitation of an extensive variety of fuels probably.

5.   Types of biofuel cells

A variety of biological fuel cells are the spotlight of dynamic investigation

  i.   Fuel cells which employ a most important fuel normally an organic dissipate e.g., green leaves which are present on the outer surface of corn cob and produce a

Materials Research Forum LLC
doi: http://dx.doi.org/10.21741/9781644900079-2

hydrogen which is afterwards exploited as a derivative fuel in a conservative oxygen or hydrogen fuel cells.

ii.   Those cells which produce voltage directly from an organic fuel e.g., glucose using enzymes or whole micro-organisms.

iii. The type of cells which merge the use of biological moieties and photo-chemically vigorous systems in the direction of isolating the power from daylight moreover change this during electrical energy [12].

5.1   Fuel cells depending on alteration of organic waste to secondary fuels

These are not called as real bio-fuel cells, however, signify a mixture of a fuel cell and bioreactor. The most attractive point of this arrangement is that it cannot just produce electrical energy, however, also uses an extensive collection of organic substances, for example, corn husks, whey or injurious waste. Fermentation can be employed to generate substrate like hydrogen or ethanol which can then be utilized for providing power to a conservative hydrogen or oxygen fuel cell [20].

5.2   Biofuel cells which directly convert fuel to electricity

Such kind of cells uses biological moieties like living cells or enzymes to produce energy directly from chemical energy which is present in organic or inorganic species. Two electrodes separated by a semipermeable membrane are dipped in a solution. Natural species for example enzyme or microbial cell can either present in solution inside the compartment of anode compartment within the cell or instead subsist in immobilized at the electrode. For one if a fine fuel is added, it either incompletely or absolutely oxidized on the anode and the electrons discharged through this method are utilized to reduce oxygen at cathode [20].

Cohen and Potter employed living cells as the energetic element. Further Davis and Yarborough in 1962 used glucose oxidase or *Escherichia coli* to half-cell having glucose which permits little current to be produced. High current could be attained by adding up methylene blue to the system. This may be illustrated through the electron flow from micro-organisms to the electrode being an incredibly ineffective practice and it follows the existence of a trouble-free mediator compound (methylene blue) which highly enhances the effectiveness of the cell. Further research involved the use of dichloro-indophenol as a mediator in a glucose oxidase-dependent cell with 100% efficiency [24].

Materials Research Forum LLC
doi: http://dx.doi.org/10.21741/9781644900079-2

5.3 Microbial biofuel cells

Microbial fuel cells were first employed to generate energy from the electrical current produced through bacteria, however; there has been a development in these systems ending in applications for other functions. High voltage was added to the potential produced through bacteria facilitating for several products (CH_4, H_2O_2 and H_2) to be produced at the cathode like methane, hydrogen peroxide and hydrogen. In microbial fuel cells, membranes can be employed to facilitate water desalination and at the same time producing electrical energy [25].

Microbial or micro-organism based fuel cells have been important as an attractive and frequently increasing discipline of science and technology which merge natural catalytic redox activity with standard abiotic electrochemical reactions and physics [26, 27]. The accumulation of biological organisms reliable for catalyzing electrochemical reactions provides these systems with a level of complication which is possibly greater than that of already complicated electrochemical systems (bio-fuel cells, super-capacitors or batteries). The major differences of microbial fuel cells with the conservative low-temperature fuel cells are

(i) at anode, the electrocatalysts is biotic [28, 29]

(ii) temperature varies between 15 to 45 °C with close to ambient levels as optimal [30]

(iii) impartial pH operational states [31, 32]

(iv) use of complicated biomass as anodic fuel [33]

(v) a promising feasible environmental impact calculated by life cycle examination [34].

There are some issues arising by using microbial fuel cell outside the lab e.g., the electrode material cost should be decreased, expensive metal cannot be employed and the latest densities should be increased. However, these problems can be solved in the laboratory, further problems remain which in the end need field testing. To solve the problem, pilot scale analysis required to analyze the functioning of these devices along with time, maintenance, temperature and changes in fuel manufacturing [25].

5.3.1 Enzyme-based biofuel cells

Three diverse enzymatic bio-anodes with (a) amino acid oxidase, (b) alcohol dehydrogenase and (c) glucose oxidase have been constructed and evaluated by Kimble and co-workers in 1964. In a comprehensive biofuel cell, they produced a positive open circuit for the two oxidases [9] and therefore such work initiated the research area called enzymatic biofuel cells. Enzymatic biofuel cells produce power under mild conditions by the oxidation of renewable energy sources avoiding greenhouse gas production or

Materials Research Foundations **44** (2019) 28-50

doi: http://dx.doi.org/10.21741/9781644900079-2

environmental pollution. Lenient operating situations such as neutral pH, renewable energy sources and ambient temperature made the enzymatic biofuel cells a capable candidate for the production of uncontaminated and proficient energy and also a substitute energy source to conservative fuel cells. There has been interesting research devoted to the design and improvement of enzymatic biofuel cells, and before they can compete with traditional fuel cells, developments in direction of totally operational enzymatic biofuel cells are needed [12, 35-38]. Before the full potential of enzymatic biofuel cells can be recognized, the properties of single bio-electrodes regarding enzyme immovability and function, an electrode material that develop electron transfer and power output and guarantee durable operation in actual environmental conditions have to be improved.

Some difficulties are present in the fabrication of enzymatic biofuel cells. The first difficulty is the maintenance of the 3D structure and enzyme catalytic active site needed to guarantee effective bio-catalytic alteration and attain high current densities [7, 38]. Secondly, guaranteed effectual electron contact among the immobilized enzyme and electrode, however, the electroactive site of mainly oxidation-reduction enzymes is curved profound in the three-dimensional structure. The enzyme-based biofuel cells power output is derived from the current flow and the rate of electron transport; both of these perform a major function in efficiency maintenance of biofuel cells. This needs to develop the enzyme immobilization and material approaches to protect and alleviate the enzyme in its steric functional situation and efficiently associate enzymes on electrode surface [37-40]. In this complicated system, the enzyme immobilization level and the stuff employed to connect the enzyme perform a major function in the performance and design of enzymatic biofuel cells.

6. Uses of biofuel cells

Here are some impending uses of biofuel cells along with those having majority attractive applications as explained below.

6.1 Transport and energy generation

The biggest resource of power in the whole world is the utilization of fossil fuels and specifically petroleum. The burning of hydrocarbons is undesirable as it causes environmental pollution. The use of cells with a power supply by means of carbohydrates would facilitate to lessen few problems. It has been estimated that one-litre concentrated solution of carbohydrate could power a car for 25 to 30 km. Coincidently, these not only provide environmental advantages, however, but it would also eliminate the danger linked with transportation of a great number of flammable fuels [20].

6.2   Wastewater treatment

Some fuel cells produce energy through the oxidation of compounds present in wastewater streams [20, 41, 42]. Through this procedure, two valuable functions have been recognized

(i) The elimination of organic compound from wastewater streams

(ii) The production of electrical power

The latest research in this area estimated that the wastewater from a town of one lac and fifty thousand people could efficiently be employed to produce more than 2.3 MW of power, however, a power of 0.5 MW might be more practical [25]. It should be described in this perspective that more than 80% of the chemical oxygen demand of wastewater can be eliminated through treatment in a microbial biofuel cell and there is a chance that electricity produced in this situation could be employed on site to power more treatment of wastewater.

6.3   Implantable power sources

Biofuel cells can efficiently be run in alive organisms, as the oxygen and fuel needed for their function can possibly subsist in use from their instantaneous atmosphere and this provides great potential as an energy source in a variety of possible insertable medical devices. For instance, a biosensor for glucose has been created using a glucose oxidase-dependent anode and cathode of cytochrome c to produce electrical power. This procedure has been employed in a biosensor design to provide a calculation of the glucose concentration within the series of 1 to 80 mM. Almost the same sensor has also been created for lactate [19]. Further efficient uses for minute bio-fuel cells consist of power supplies for drug distribution systems by way of biofuel cell being tiny as much as necessary for this process [14, 43].

7.   Use of enzymes in a different type of biofuel cells

EBFCs have difficulties related to the delicate three-dimensional structures of proteins and should be constant for catalytic activity. However, few oxidoreductases such as peroxidases and multicopper oxidases for the bio-cathode and glucose oxidases for bio-anode have better stability in solution. In many dehydrogenase dependent biofuel cells, enzymes in solution failed to retain their three-dimensional structures or get denatured during the period of 8 h to 3 days. Now a day's researchers are concentrated on enhancing the lifetime and density of energy through the use of enzymatic biofuel cells, developing electron flow pathway and innovative immobilization methods for enhancing the stability of enzyme at the electrode surface [44].

Enzymes did not provide electrons as effortlessly as metal catalysts do, which means that producing a flow of electricity from enzymes is very difficult. To solve this problem enzyme can be prepared to confer their electrons with mediators although employing mediators can produce additional troubles in the biofuel cells. An enzyme immobilizing process has been useful. In this process for proper functioning, enzymes have to be placed at a specific location where they stay in a fuel cell because animals enzymes are moving freely in a cell. Moreover, enzymes that exist naturally have much lower stability. An enzyme should have to last for months or years before the requirement of its replacement, in addition, to be efficient in an automobile or laptop. However a normal enzyme duration in the living body is only days [45].

7.1   Immobilization and stabilization of enzymes for bio-electrochemical system

Suitable blueprint of the enzyme-electrode boundary is a major concern for the synthesis of the innovative bio-electrochemical systems such as biosensors or biofuel cells. These systems usually include complicated organizations of immobilization polymers, enzymes and redox negotiators that should cooperate among the substrate of the electrode. At the University of Columbia, scientists are employing protein engineering to design multifunctional proteins and peptides which can improve and simplify the enzymatic electrode. Consequently, they are making proteins which can self-assemble into bioactive hydrogels along with the activities of redox enzymes. This removes the requirement for the integration of polymers in the system. Furthermore, peptides are being engineered which can bind the redox mediators. In the future, this molecular engineering method will spectacularly simplify the characterization, fabrication and reproducibility of bio-electro catalytic edge.

7.2   Electron transport in biofuel cells

The major problem in biofuel cells is the catalysis of electrode processes. Fuel oxidation e.g., glucose catalyzed via enzymes like glucose oxidases possibly completed straightforwardly or may include a redox mediator. The direct enzymatic oxidation needs an active site of the enzyme corresponds instantaneously with the surface of the electrode. In support of the most suitable glucose oxidase, this appeared to be quite impractical as its flavin type active site is covered very deeply in a protein shell of the pretty big enzyme molecule. To deal with this problem, Sandia National Laboratories genetically distorted enzyme to formulate its active site more accessible for linkage with the electrode. Many mutant proteins can easily be prepared in one day. The main part of the problem after preparation is recognizing which protein will function properly [45].

Materials Research Forum LLC

doi: http://dx.doi.org/10.21741/9781644900079-2

At the New Mexico University different scientists are working on the problem of electrode and enzymes in diverse angle. Not only are the enzymes important but many other things play a vital role in it including the use of the electrode in different ways. A nanostructured material particularly carbon nanotubes (CNTs) could be used. In comparison to the enzyme, CNTs are very minute in size, directly related to the active site and can play a better role. Undeviated oxidation of glucose via glucose oxidase fixed in carbon nano-tube with modified porous matrix revealed that anode electrode can be made functional at almost 400 mV [46]. Direct reduction of oxygen catalyzed by copper having enzymes e.g., bilirubin oxidases, ascorbate oxidases or laccases at a potential only 50 mV less than the thermodynamic value [47], and merging this with thermodynamic system provides the opportunity of creating a biofuel cell with such elevated open circuit current like 1 V. In order to perform all this, many things require to be suitably tuned, the enzyme surface interface especially. Enzyme orientation and charge transfer are of significance in direct electron flow. Simple engineering of the material is also possible, to permit us to get as much power as possible [45].

7.3 Complete oxidation of biofuel

Among major issues, plaguing enzymatic biofuel cells has been little energy densities because of partial oxidation of biological fuels at the anode. There is only one example of complete oxidation i.e., methanol oxidation in an enzymatic biofuel cell employing aldehyde dehydrogenase, formate dehydrogenase or alcohol dehydrogenase. With complicated biofuels e.g., sucrose and glucose, the enzymatic metabolic pathways are more complicated and both have oxidoreductase enzymes and other enzymes liable for chemical transformations. The glycolysis and Krebs cycle are the cellular pathways liable for the enzymatic breakdown of sugars. Impersonating these cellular pathways at an electrode needs the capability to immobilize the enzymes inadequate microenvironment. At the University of Saint Louis, researchers are implementing click chemistry to make enzyme complexes for reduction of transportation restrictions and improving immobilization membranes of the enzyme for protecting the suitable microenvironment.

From immediate resources, nature has developed complicated metabolic complexes to isolate chemical energy. Not a distinct nodule in the complex can symbolize the main blockage because the enzymes involved in these systems have evolved to show a dispersed power over the fluctuation of substances. Lately, in the bioelectrochemical field there has been stimulating progress and for biofuel cells and biosensors multi-enzyme systems are being made. However, these systems are frequently made by employing enzymes from diverse organisms that did not develop to function simultaneously and are not boosted to perform *in vitro* circumstances. Where the kinetic control of the system is

Materials Research Forum LLC
doi: http://dx.doi.org/10.21741/9781644900079-2

not well dispersed, this ended in the synthesis of metabolic pathways and just one node can take over the entire function. Researchers at Columbia University are implementing metabolic control examination to understand and enhance the kinetic function of biofuel cells. These perceptions can be employed to guarantee the most favourable working states in addition to make the selection of enzymes that should be employed in these synthetic metabolic complexes.

7.4 Use of bilirubin oxidases for the elaboration of the electrode in biofuel cells

There are three main conditions e.g., little sensitivity to chloride ions, high activity at pH 7 and good thermal stability for the expansion of an effective enzymatic cathode. This regard, bilirubin oxidases are interesting aspirants for the enzymatic cathode in biofuel cells since they acquire nearly all of the properties. Reiss recognized a novel CotA enzyme bacterial laccase and characterized it from *Bacillus pumilus* [48]. It has been revealed that because of its high catalytic activity with bilirubin like substrate, this enzyme should be characterized as a bilirubin oxidase [49]. Besides the high current density, to be employed as an effective enzyme cathode the bilirubin oxidase should be thermostable at neutral pH. CotA protein normally showed the greater thermostability between multi-copper oxidase with half-life >100 min at 80 °C and hypothesized that it was associated with the more dense configuration of protein or to coppers intensely covered in the enzymes. In this hypothesis, the novel bilirubin oxidase is boiled for thirty minutes to completely unfold for SDS-PAGE. As well as, it shows that the innovative bilirubin oxidase is more stable at physiological temperatures. These investigations demonstrate that this novel bilirubin oxidase is a capable enzyme for the expansion of biofuel cells. No effort was made for enhancement of the immovability of fixed enzymes via using nanotube fibers except of naked glassy carbon electrode [49, 50].

7.5 Biofuel cells based on laccase biocathodes

A major problem in employing laccases for enzymatic biofuel cell lies in developing its stability in direction of chloride [51] and reducing the hydroxyl inhibition [52]. In the presence of chloride and its reduction in activity at pH more than 5, many methods have been examined to avoid the hindrance of mostly employed laccases. Resistance to chloride can be increased by the orientation of laccases or its direct electron transport with the electrode from organisms like *Trametes versicolor* [53] and *Trametes hirsute* [52-55]. Such an effect is because of the fact that chloride ion competes with the redox reaction with a redox mediator. In view of a particular orientation, laccase on electrode surface possibly will produce steric obstructions for reaching the active site; direct electron transfer is not as much affected by the attachment of chloride to the laccase active site. Taking the benefit of a natural multiplicity of laccase sources and to function

Enzymatic Fuel Cells   Materials Research Forum LLC
Materials Research Foundations **44** (2019) 28-50   doi: http://dx.doi.org/10.21741/9781644900079-2

at physiological pH, laccases from other organisms were also analyzed. Directed mutagenesis has been confirmed to design an efficient source of laccase mutant for functioning at an elevated potential, impartial pH and in the prevalence of chloride [52]. A local pH control method has been proposed by Clot *et al.*, [56]. Laccases were combined with glucose oxidase in the bi-enzymatic system. Glucose oxidase-generated gluconic acid at the surrounding area of laccases, reducing the local pH and increasing electro-catalytic oxygen reduction in the neutral pH solution.

Recently, for laccases immobilized in osmium hydro-gels, the main function of hydrogen peroxide in the bioelectrocatalytic reduction of oxygen through laccases have been reported [57]. Throughout electrocatalytic reduction of oxygen into water, immobilized laccase also has an incomplete reduction of oxygen to hydrogen peroxide, which causes inhibitory effects for laccase. Additionally, bioanodes is chiefly depending on glucose oxidase which shows a feasible H_2O_2 source. In fact, not all glucose oxidase enzymes are wired electrically and therefore, a portion of the enzymes catalyzes the oxidation of glucose with associated production of H_2O_2 in the presence of oxygen. Milton and coworkers have confirmed the potential inhibition of the laccase dependent bio-cathodes via H_2O_2 generated at the bio-anode [58]. Like a substitute, Milton and coworker have projected the alternate of glucose oxidase by a FAD-glucose dehydrogenase [59].

7.6 *In vivo* use of tyrosinase and glucose oxidase in glucose-based biofuel cells

Cinquin and coworkers [15], surgically set a glucose-based biofuel cell in the abdomen of a rat in 2010. This glucose based biofuel cells depend on MET (mediated electron transport) utilizing glucose oxidases at the anode and tyrosinase at the cathode. Tyrosinase causes the reduction of oxygen, as this enzyme confirms its optimum activity at neutral pH and is not inhibited by chloride. Furthermore, for dopamine detection, this enzyme was incorporated in the brain of a rat like an amperometric biosensor. Conversely, this glucose based biofuel cells transported low power output of 6l W and low voltage of 0.25 V. The enormous efforts in the development of electron transfer between enzymes and electrodes, particularly for laccases, have provided access to higher voltages and higher power outputs, enough to increase linked boost converters for power management and consequent power delivery of small electronic devices. Glucose-based biofuel cells in living organisms inserted by Szczupak and coworkers [60] did not need difficult surgery like clams, snails [61] and lobsters [62]. Their glucose based biofuel cells were dependent on the direct wiring of pyrroloquinolinequinone glucose dehydrogenase (PQQ-GDH) for oxidation of glucose and laccase for oxygen reduction. Both enzymes were bounded to carbon nanotubes-dependent papers through an activated ester-modified pyrene. Through linking some lobsters in sequence, a cell voltage of 1 V

Materials Research Forum LLC
doi: http://dx.doi.org/10.21741/9781644900079-2

could be attained for powering a watch. Five glucose based biofuel cells were also attached in succession and were capable to contribute *in vitro* a cardiac pacemaker from physiological serum. They also placed their elastic bio-electrodes on the surgically uncovered rat cremaster tissue [59, 63].

Conclusion

As summarized above enzymes are used in many industrial processes and new areas of application are continuously being added. Developments in biotechnology, protein engineering and recombinant technology have evolved the enzymes to such extent where no one would have thought an enzyme could be applicable. The introduction of the enzyme as a catalyst in biofuel cells results in significant conversion and storage of energy. In a world of increasing population, this enzyme technology provides a great potential to overcome different challenges related to natural resources depletion for energy production.

References

[1]   T. Enzym, Ueber das Verhalten verschiedener organisirtler und sog. ungeformter, 62 (1976) 3–7.

[2]   N. Gurung, S. Ray, S. Bose, V. Rai, W.F. K, A broader view : Microbial enzymes and their relevance in industries, medicine, and beyond BioMed Research International (2013) Article ID 329121, 18 pages.

[3]   R. Vallery, R.L. Devonshire, Life of Pasteur, (2003).

[4]   A. Ullmann, Pasteur–Koch, distinctive ways of thinking about infectious diseases, Microbe 2 (2007) 383–387.

[5]   H. Nadeem, M. Hamid, M. Hussnain, F. Azeem, S. Muzammil, M. Rizwan, M. Amjad, I. Rasul, M. Riaz, Microbial invertases : A review on kinetics, thermodynamics, physiochemical properties, Process Biochem. 50 (2015) 1202–1210. https://doi.org/10.1016/j.procbio.2015.04.015

[6]   A. Bairoch, C.M. Universitaire, M. Servet, The ENZYME database in 2000, Nucleic Acids Res. 28 (2000) 304–305. https://doi.org/10.1093/nar/28.1.304

[7]   M.J. Cooney, V. Svoboda, C. Lau, G. Martin, S.D. Minteer, Enzyme catalysed biofuel cells, Energy Environ. Sci. 1 (2008) 320–337. https://doi.org/10.1039/b809009b

Materials Research Forum LLC
doi: http://dx.doi.org/10.21741/9781644900079-2

[8]   Z. Ghassemi, G. Slaughter, Biological fuel cells and membranes, Membranes (Basel). 7 (2017) 3. https://doi.org/10.3390/membranes7010003

[9]   A.T. Yahiro, S.M. Lee, D.O. Kimble, Bioelectrochemistry: I. Enzyme utilizing bio-fuel cell studies, Biochim. Biophys. Acta (BBA)-Specialized Sect. Biophys. Subj. 88 (1964) 375–383.

[10]   Y.F. Choo, J. Lee, I.S. Chang, B.H. Kim, Bacteria communities in microbial fuel cells enriched with high concentrations of glucose and glutamate, J. Microbiol. Biotechnol. 16 (2006) 1481–1484.

[11]   S. Topcagic, S.D. Minteer, Development of a membraneless ethanol/oxygen biofuel cell, Electrochim. Acta 51 (2006) 2168–2172. https://doi.org/10.1016/j.electacta.2005.03.090

[12]   M.T. Meredith, S.D. Minteer, Biofuel cells : Enhanced enzymatic bioelectrocatalysis, Annual Review of Anal. Chem. 5 (2012) 157-179. https://doi.org/10.1146/annurev-anchem-062011-143049

[13]   A. Heller, B. Feldman, Electrochemical glucose sensors and their applications in diabetes management, Chem. Rev. (2008) 2482–2505. https://doi.org/10.1021/cr068069y

[14]   N. Mano, F. Mao, A. Heller, Characteristics of a miniature compartment-less glucose– O2 biofuel cell and its operation in a living plant, J. Am. Chem. Soc. 125 (2003) 6588–6594. https://doi.org/10.1021/ja0346328

[15]   P. Cinquin, C. Gondran, F. Giroud, S. Mazabrard, A. Pellissier, F. Boucher, J.-P. Alcaraz, K. Gorgy, F. Lenouvel, S. Mathé, A glucose biofuel cell implanted in rats, PLoS One 5 (2010) e10476. https://doi.org/10.1371/journal.pone.0010476

[16]   S. Fishilevich, L. Amir, Y. Fridman, A. Aharoni, L. Alfonta, Surface display of redox enzymes in microbial fuel cells, J. Am. Chem. Soc. 131 (2009) 12052–12053. https://doi.org/10.1021/ja9042017

[17]   C.M. Moore, N.L. Akers, A.D. Hill, Z.C. Johnson, S.D. Minteer, Improving the environment for immobilized dehydrogenase enzymes by modifying Nafion with tetraalkylammonium bromides, Biomacromolecules 5 (2004) 1241–1247. https://doi.org/10.1021/bm0345256

[18]   E. Katz, A.F. Bückmann, I. Willner, Self-powered enzyme-based biosensors, J. Am. Chem. Soc. 123 (2001) 10752–10753. https://doi.org/10.1021/ja0167102

[19]   F. Davis, P.J. Higson, Biofuel cells — Recent advances and applications, Biosens. Bioelect. 22 (2007) 1224–1235. https://doi.org/10.1016/j.bios.2006.04.029

Materials Research Forum LLC
doi: http://dx.doi.org/10.21741/9781644900079-2

[20] G.T.R. Palmore, G.M. Whitesides, Microbial and enzymatic biofuel cells, in: ACS Publications, 1994. https://doi.org/10.1021/bk-1994-0566.ch014

[21] A. Mitsos, I. Palou-Rivera, P.I. Barton, Alternatives for micropower generation processes, Ind. Eng. Chem. Res. 43 (2004) 74–84. https://doi.org/10.1021/ie0304917

[22] M. Doyle, G. Rajendran, W. Vielstich, H.A. Gasteiger, A. Lamm, Handbook of Fuel Cells Fundamentals, Technology and Applications, Fuel Cell Technol. Appl. 3 (2003).

[23] M.K. Weibel, C. Dodge, Biochemical fuel cells: Demonstration of an obligatory pathway involving an external circuit for the enzymatically catalyzed aerobic oxidation of glucose, Arch. Biochem. Biophys. 169 (1975) 146–151. https://doi.org/10.1016/0003-9861(75)90327-6

[24] B.E. Logan, K. Rabaey, Conversion of wastes into bioelectricity and chemicals by using microbial electrochemical technologies, Science 337 (2012) 686–690. https://doi.org/10.1126/science.1217412

[25] B.E. Logan, M. Elimelech, Membrane-based processes for sustainable power generation using water, Nature 488 (2012) 313-319. https://doi.org/10.1038/nature11477

[26] A.P. Borole, G. Reguera, B. Ringeisen, Z.-W. Wang, Y. Feng, B.H. Kim, Electroactive biofilms: Current status and future research needs, Energy Environ. Sci. 4 (2011) 4813–4834. https://doi.org/10.1039/c1ee02511b

[27] U. Schroeder, F. Harnisch, Biofilms, Electroactive, in: Encycl. Appl. Electrochem., Springer, 2014: pp. 120–126. https://doi.org/10.1007/978-1-4419-6996-5_249

[28] P.-F. Tee, M.O. Abdullah, I.A.W. Tan, M.A.M. Amin, C. Nolasco-Hipolito, K. Bujang, Effects of temperature on wastewater treatment in an affordable microbial fuel cell-adsorption hybrid system, J. Environ. Chem. Eng. 5 (2017) 178–188. https://doi.org/10.1016/j.jece.2016.11.040

[29] Y. Ahn, B.E. Logan, Saline catholytes as alternatives to phosphate buffers in microbial fuel cells, Bioresour. Technol. 132 (2013) 436–439. https://doi.org/10.1016/j.biortech.2013.01.113

[30] Y. Ye, X. Zhu, B.E. Logan, Effect of buffer charge on performance of air-cathodes used in microbial fuel cells, Electrochim. Acta. 194 (2016) 441–447. https://doi.org/10.1016/j.electacta.2016.02.095

Enzymatic Fuel Cells
Materials Research Foundations **44** (2019) 28-50

Materials Research Forum LLC
doi: http://dx.doi.org/10.21741/9781644900079-2

[31] P. Pandey, V.N. Shinde, R.L. Deopurkar, S.P. Kale, S.A. Patil, D. Pant, Recent advances in the use of different substrates in microbial fuel cells toward wastewater treatment and simultaneous energy recovery, Appl. Energy. 168 (2016) 706–723. https://doi.org/10.1016/j.apenergy.2016.01.056

[32] D. Pant, A. Singh, G. Van Bogaert, Y.A. Gallego, L. Diels, K. Vanbroekhoven, An introduction to the life cycle assessment (LCA) of bioelectrochemical systems (BES) for sustainable energy and product generation: relevance and key aspects, Renew. Sustain. Energy Rev. 15 (2011) 1305–1313. https://doi.org/10.1016/j.rser.2010.10.005

[33] D. Leech, P. Kavanagh, W. Schuhmann, Enzymatic fuel cells: Recent progress, Electrochim. Acta. 84 (2012) 223–234. https://doi.org/10.1016/j.electacta.2012.02.087

[34] M.C. Beilke, T.L. Klotzbach, B.L. Treu, D. Sokic-Lazic, J. Wildrick, E.R. Amend, L.M. Gebhart, R.L. Arechederra, M.N. Germain, M.J. Moehlenbrock, Enzymatic Biofuel Cells, in: Micro Fuel Cells, Elsevier, 2009: pp. 179–241. https://doi.org/10.1016/B978-0-12-374713-6.00005-6

[35] S. Aquino Neto, A.R. De Andrade, New energy sources: The enzymatic biofuel cell, J. Braz. Chem. Soc. 24 (2013) 1891–1912. https://doi.org/10.5935/0103-5053.20130261

[36] A. Karimi, A. Othman, A. Uzunoglu, L. Stanciu, S. Andreescu, Graphene based enzymatic bioelectrodes and biofuel cells, Nanoscale. 7 (2015) 6909–6923. https://doi.org/10.1039/C4NR07586B

[37] X.-Y. Yang, G. Tian, N. Jiang, B.-L. Su, Immobilization technology: a sustainable solution for biofuel cell design, Energy Environ. Sci. 5 (2012) 5540–5563. https://doi.org/10.1039/C1EE02391H

[38] K. Rabaey, N. Boon, S.D. Siciliano, M. Verhaege, W. Verstraete, Biofuel cells select for microbial consortia that self-mediate electron transfer, Appl. Environ. Microbiol. 70 (2004) 5373–5382. https://doi.org/10.1128/AEM.70.9.5373-5382.2004

[39] G.-C. Gil, I.-S. Chang, B.H. Kim, M. Kim, J.-K. Jang, H.S. Park, H.J. Kim, Operational parameters affecting the performannce of a mediator-less microbial fuel cell, Biosens. Bioelectron. 18 (2003) 327–334. https://doi.org/10.1016/S0956-5663(02)00110-0

Materials Research Forum LLC
doi: http://dx.doi.org/10.21741/9781644900079-2

[40]   C.M. Moore, S.D. Minteer, R.S. Martin, Microchip-based ethanol/oxygen biofuel cell, Lab Chip. 5 (2005) 218–225. https://doi.org/10.1039/b412719f

[41]   S.D. Minteer, B.Y. Liaw, M.J. Cooney, Enzyme-based biofuel cells, Curr. Opin. Biotechnol. 18 (2007) 228–234. https://doi.org/10.1016/j.copbio.2007.03.007

[42]   P. Atanassov, C. Apblett, S. Banta, S. Brozik, S.C. Barton, M. Cooney, B.Y. Liaw, S. Mukerjee, S.D. Minteer, Enzymatic biofuel cells, Interface-Electrochemical Soc. 16 (2007) 28–31.

[43]   D. Ivnitski, B. Branch, P. Atanassov, C. Apblett, Glucose oxidase anode for biofuel cell based on direct electron transfer, Electrochem. Commun. 8 (2006) 1204–1210. https://doi.org/10.1016/j.elecom.2006.05.024

[44]   G. Gupta, V. Rajendran, P. Atanassov, Bioelectrocatalysis of oxygen reduction reaction by laccase on gold electrodes, Electroanal. An Int. J. Devoted to Fundam. Pract. Asp. Electroanal. 16 (2004) 1182–1185. https://doi.org/10.1002/elan.200403010

[45]   P. Atanassov, C. Apblett, S. Banta, S. Brozik, S.C. Barton, M. Cooney, B.Y. Liaw, S. Mukerjee, S.D. Minteer, Enzymatic biofuel cells, The Electrochemical Society *Interface* (2007).

[46]   R. Reiss, J. Ihssen, L. Thöny-Meyer, Bacillus pumilus laccase: a heat stable enzyme with a wide substrate spectrum, BMC Biotechnol. 11 (2011) 9. https://doi.org/10.1186/1472-6750-11-9

[47]   F. Durand, C.H. Kjaergaard, E. Suraniti, S. Gounel, R.G. Hadt, E.I. Solomon, N. Mano, Bilirubin oxidase from Bacillus pumilus: a promising enzyme for the elaboration of efficient cathodes in biofuel cells, Biosens. Bioelectron. 35 (2012) 140–146. https://doi.org/10.1016/j.bios.2012.02.033

[48]   F. Gao, L. Viry, M. Maugey, P. Poulin, N. Mano, Engineering hybrid nanotube wires for high-power biofuel cells, Nat. Commun. 1 (2010) 2-7. https://doi.org/10.1038/ncomms1000

[49]   C. Vaz-Dominguez, S. Campuzano, O. Rüdiger, M. Pita, M. Gorbacheva, S. Shleev, V.M. Fernandez, A.L. De Lacey, Laccase electrode for direct electrocatalytic reduction of O_2 to H_2O with high-operational stability and resistance to chloride inhibition, Biosens. Bioelectron. 24 (2008) 531–537. https://doi.org/10.1016/j.bios.2008.05.002

[50] D.M. Mate, D. Gonzalez-Perez, M. Falk, R. Kittl, M. Pita, A.L. De Lacey, R. Ludwig, S. Shleev, M. Alcalde, Blood tolerant laccase by directed evolution, Chem. Biol. 20 (2013) 223–231. https://doi.org/10.1016/j.chembiol.2013.01.001

[51] N. Lalaoui, K. Elouarzaki, A. Le Goff, M. Holzinger, S. Cosnier, Efficient direct oxygen reduction by laccases attached and oriented on pyrene-functionalized polypyrrole/carbon nanotube electrodes, Chem. Commun. 49 (2013) 9281–9283. https://doi.org/10.1039/c3cc44994g

[52] C. Gutiérrez-Sánchez, M. Pita, C. Vaz-Dominguez, S. Shleev, A.L. De Lacey, Gold nanoparticles as electronic bridges for laccase-based biocathodes, J. Am. Chem. Soc. 134 (2012) 17212–17220. https://doi.org/10.1021/ja307308j

[53] S. Clot, C. Gutierrez-Sanchez, S. Shleev, A.L. De Lacey, M. Pita, Laccase cathode approaches to physiological conditions by local pH acidification, Electrochem. Commun. 18 (2012) 37–40. https://doi.org/10.1016/j.elecom.2012.01.022

[54] P. Scodeller, R. Carballo, R. Szamocki, L. Levin, F. Forchiassin, E.J. Calvo, Layer-by-layer self-assembled osmium polymer-mediated laccase oxygen cathodes for biofuel cells: the role of hydrogen peroxide, J. Am. Chem. Soc. 132 (2010) 11132–11140. https://doi.org/10.1021/ja1020487

[55] R.D. Milton, F. Giroud, A.E. Thumser, S.D. Minteer, R.C.T. Slade, Hydrogen peroxide produced by glucose oxidase affects the performance of laccase cathodes in glucose/oxygen fuel cells: FAD-dependent glucose dehydrogenase as a replacement, Phys. Chem. Chem. Phys. 15 (2013) 19371–19379. https://doi.org/10.1039/c3cp53351d

[56] A. Le Goff, M. Holzinger, S. Cosnier, Recent progress in oxygen-reducing laccase biocathodes for enzymatic biofuel cells, Cell. Mol. Life Sci. 72 (2015) 941–952. https://doi.org/10.1007/s00018-014-1828-4

[57] A. Szczupak, J. Halámek, L. Halámková, V. Bocharova, L. Alfonta, E. Katz, Living battery–biofuel cells operating in vivo in clams, Energy Environ. Sci. 5 (2012) 8891–8895. https://doi.org/10.1039/c2ee21626d

[58] L. Halámková, J. Halámek, V. Bocharova, A. Szczupak, L. Alfonta, E. Katz, Implanted biofuel cell operating in a living snail, J. Am. Chem. Soc. 134 (2012) 5040–5043. https://doi.org/10.1021/ja211714w

[59] K. MacVittie, J. Halámek, L. Halámková, M. Southcott, W.D. Jemison, R. Lobel, E. Katz, From "cyborg" lobsters to a pacemaker powered by implantable biofuel cells, Energy Environ. Sci. 6 (2013) 81–86. https://doi.org/10.1039/C2EE23209J

Materials Research Forum LLC
doi: http://dx.doi.org/10.21741/9781644900079-2
[60] A. Le, G. Michael, H. Serge, Recent progress in oxygen-reducing laccase biocathodes for enzymatic biofuel cells, Cell. Mol. Life Sci. (2015) 941–952.

[61] J.A. Castorena-Gonzalez, C. Foote, K. MacVittie, J. Halámek, L. Halámková, L.A. Martinez-Lemus, E. Katz, Biofuel cell operating in vivo in rat, Electroanalysis. 25 (2013) 1579–1584. https://doi.org/10.1002/elan.201300136

Enzymatic Fuel Cells
Materials Research Foundations **44** (2019) 51-72

Materials Research Forum LLC
doi: http://dx.doi.org/10.21741/9781644900079-3

Chapter 3

Current Advances in Immobilization Techniques of Enzymes

Rajeev Ravindran and Amit K. Jaiswal*

School of Food Science and Environmental Health, College of Sciences and Health, Technological University Dublin, Cathal Brugha Street, Dublin 1, Republic of Ireland

amit.jaiswal@dit.ie; akjaiswal@outlook.com

Abstract

Enzymes are biological catalysts that accelerate the rate of a reaction without themselves being consumed. They sustain their activity for long periods of time and therefore are widely used in industrial processes. Enzymes contribute to approximately 28% of the operating cost of a production process; however, in most of the cases, the enzymes involved in production processes cannot be retrieved. Enzyme immobilisation is the process of attaching an enzyme molecule to a solid support with the intentions of its reuse, production and purification. Proper immobilisation of an enzyme on a support is dependent on the properties of the enzyme and carrier material. The binding to a support material can be temporary or permanent depending upon the chemical bond formed between the enzyme and the support. This article discusses in detail the intrinsic factors that influence enzyme immobilisation and the latest techniques such as one-step immobilisation, microencapsulation and cross-linked aggregates that have been proposed in recent years.

Keywords

Enzyme Immobilization, Carrier Material, Entrapment, Affinity Tags, Adsorption, One Step Immobilization

Contents

Materials Research Forum LLC
doi: http://dx.doi.org/10.21741/9781644900079-3

1. Introduction

Enzymes are biological catalysts that facilitate reactions to take place within living systems. They play an active role in biological processes and therefore their influence on health and diseases have been actively researched upon. They have been used for various applications, especially in food processing. Enzymes under mild conditions can catalyse chemical reactions at very high rates with a high degree of substrate specificity eliminating the formation of by-products. They display high region, chemo and enantioselectivities while operating under mild conditions [1,2]. Extensive research on enzymes and their properties in the early 20th century led to the knowledge of relevant information about their structure, chemical composition and modes of action. This resulted in the widespread technological use of enzymes in several fields such as textile, pharmaceutical and food industries. However, there are certain problems related to the applications of enzymes in industrial processes [3-7].

Enzymes once suspended in an aqueous reaction medium are almost impossible to retrieve or recycle. Additionally, due to their proteinaceous nature enzymes are highly unstable and require an aqueous environment to catalyse reactions [8]. Enzymes can catalyse reactions in different physical states: as individual molecules, in the collection with other moieties, and fastened to surfaces. This unique property of enzymes is useful while 'immobilising enzymes'. Enzyme immobilisation is the process of attaching an enzyme molecule to a solid support with the intentions of its reuse, production and purification without loss in their catalytic activity thereby dramatically improving process economy. It also enables the use of insoluble enzymes suspended in hydrophobic organic media by optimising the enzyme dispersion and improving accessibility towards the substrate without the aggregation of hydrophilic protein particles [9]. Although theoretically promising, practical experiences of enzyme immobilisation have not always been fruitful as per expectations. Substantial design approaches and parameters need to be considered while selecting the mode of immobilisation. This book chapter is aimed to review the different parameters that need to be considered for enzyme immobilisation along with the latest technologies associated with it.

2.   Factors to consider prior to enzyme immobilization

Enzyme immobilisation essentially involves two players: the carrier material and the enzyme itself. Enzymes being proteins will have polar regions contributed by the presence of charged groups on amino acids such as lysine and glutamic acid and apolar regions that may comprise of sugar moieties. The presence of charged species and hydrophobic components can influence the surface properties of the enzyme molecule [10]. Therefore, the carrier material should be designed keeping in mind the presence of both regions on the enzyme. Carrier materials can be classified into organic and inorganic according to their chemical composition. Organic enzyme support materials can be categorized further as natural and synthetic according to their mode of origin. Carrier materials are required to have a high surface area, that can be achieved either by choosing particles with smaller sizes or porous materials with pores of sufficiently large diameters so as to not hinder the diffusion of substrates. This is particularly important in the case of enzyme encapsulation where the enzyme is locked inside the carrier and efficient catalysis is dependent upon the substrate diffusion rate [11]. The choice of carrier material for a particular enzyme molecule is really a question of trial and error along with careful analysis of surface properties and functional groups. The same can be said about the choice of the immobilisation method. The most important factor that needs be taken under consideration is that the enzyme should be stable during and after immobilisation [12].

Although one may expect a dip in the activity of the enzyme post immobilisation compared to the native one it is not always the case. Conducting a detailed study based on the different parameters of the enzyme and the carrier can lead to the formation of enzyme preparations that can perform better than the native enzyme. Usually, three terms are used to determine the success of immobilisation: immobilisation yield, immobilisation efficiency and activity recovery.

Immobilisation yield is defined as the percentage ratio of immobilised activity versus the starting activity. It is used to measure the extent of immobilisation.

Yield (%) = 100 x (immobilised activity/starting activity)

Immobilised activity is determined by the total residual enzyme activity that remains in the solution after the immobilisation and then subtracting this number from the total starting activity. Running a free blank experiment parallel to the immobilisation process helps in compensating free enzyme deactivation under immobilisation conditions.

Immobilisation efficiency may be defined as the percentage of bound enzyme activity that is present in the immobilisation.

Immobilisation efficiency (%) = 100 x (observed activity/immobilisation activity).

Theoretically, it is possible to achieve 100% yield efficiency and no observable activity post immobilisation as the process would have deactivated the enzyme or become inaccessible. Meanwhile, activity recovery is a direct indication of the success of the immobilisation process. It is a product of the immobilisation yield and the immobilisation efficiency.

Activity recovery (%) = (observed activity/starting activity)

All the above terms need to be assessed in terms of total activities (μmol min^{-1}) and not by using specific activities (μmol min^{-1}mg^{-1}). Also, the exact same assay conditions need to be used to measure all activities. The comparison of the observed immobilised activity with free enzyme hugely depends upon the activity assay, which includes properties of the substrate, pH and temperature and the nature of the biocatalyst involved. An enzyme immobilised in a polar matrix will exhibit a higher observed activity in the presence of a polar substrate which has a small particle size at higher concentrations than substrates that are apolar and low in concentration [13].

The physical and chemical properties of the carrier materials are of utmost importance in determining their viability in enzyme immobilisation. While a lot of factors contribute to a matrix being employed to encapsulate enzymes, the cost is one of the major deciding factors. The matrix chosen for enzyme immobilisation should ideally be affordable, eco-friendly as well as easily available. It should be inert to any chemical reactions and should not interfere with the enzymatic process. It should be stable over a wide temperature and pH range while protecting the enzyme attached. Regeneration is another desirable quality of an enzyme immobilisation carrier where it can be revived after its

Materials Research Forum LLC
doi: http://dx.doi.org/10.21741/9781644900079-3

useful lifetime. A carrier material that can pack a large amount of enzyme is highly desirable. A reduced hydrophobicity of the carrier will assure the desorption of undesired proteins onto the matrix. Furthermore, nonspecific adsorption antimicrobial property is the additional desired quality of carrier material. As obvious, all these properties cannot be observed in any single type of enzyme immobilisation matrix. Table 1 represents some of the common matrices used for the immobilisation of different enzymes.

Table 1 Enzyme support materials and their properties

Carrier material	Properties	Enzyme immobilised	Reference
Poly(3-hydroxybutyrate-co-hydroxyvalerate)	Biocompatibility, biodegradability, strength, easy reabsorption, non-toxicity, eco-friendliness	Candida rugosa lipase	[14]
Zeolite	Molecular sieves, microporous, crystalline	Lysosyme, papain, trypsin	[15, 16]
Hydrotalcite	Anion exchange properties	Lipozyme-TL IM	[17]
Mesoporous silica	Highly porous, adsorption, immobilization influenced by electrostatic interaction	Lysosyme	[18]
Gold nanoparticles	Controllable particle size, structure, non-toxic, hydrophobic/hydrophilic and biocompatibile	Trypsin	[19]
Chitosan	Inexpensive, natural polyaminosaccharide, non-toxic, biocompatible and biodegradable	β-Galactosidase	[20]
Paper-based carriers	Inexpensive, non-toxic, biocompatible and biodegradable, abundantly available	Glucose oxidase biosensor	[21]
Collagen	Formation of the porous structure, permeabile, hydrophilic and biocompatibile	Alkaline phosphatase	[22]
Alginate/silica biocomposites	Inorganic porous network, control over diffusion properties	β-galactosidase	[23]

3.   Binding to a support

The choice of best immobilisation method for a particular enzyme should be determined on the basis of certain characters that are exclusive to both enzyme and carrier material respectively. Generally, immobilisation techniques can be categorised under four titles with each category having their respective advantages and disadvantages.

3.1   Non-covalent adsorption, deposition and hydrogen bonds

Adsorption is one of the most popular methods used for enzyme immobilisation. Physical interactions between enzyme and carrier including weak forces such as van der Waal's forces, ionic interactions, entropy changes and hydrogen bonding hold both the components together. A case where van der Waal's forces and entropy changes play an active role in immobilisation is when an enzyme with lipophilic surface area comes in contact with a hydrophobic carrier. On the other hand, in the case of a hydrophilic enzyme, employing a hydrophilic carrier ensures the formation of hydrogen bonding between the two components. Although the bonding strength is weak, and this has actually been an advantage as it does not distort the confirmation of the enzyme and thus maintaining the active site intact and retaining its activity. Theoretically, any carrier qualifies for enzyme adsorption albeit the choice of the enzyme depends upon the presence of active functional groups on the carrier. Enzyme-carrier affinity is the key to this type of immobilisation and an absence of functional groups can be overcome by the introduction of intermediate agents such as carrier modifiers. There are numerous choices available for a carrier material if adsorption is the chosen method of enzyme immobilisation. However, factors such as cost, availability, stability under defined conditions and the type of reactors need to be taken into account while selecting the perfect carrier substance [24]. One of the main disadvantages of this technique is the leaching of the enzyme from the carrier material in an aqueous reaction mixture. This is however not the case when employing adsorption based immobilised enzymes in organic solvents. The intrinsic insolubility of enzymes in organic solvents ensures that enzymes remain intact in the carrier material [25]. In the case of enzyme immobilisation via deposition, the carrier is simply added into the aqueous mixture of enzymes. The biocatalyst is recovered by simple precipitation or evaporation of the aqueous phase. This leads to the enzyme being 'deposited' on the support media without adsorption or entropy changes to hold it together. Immobilisation techniques using celite powder employ this strategy [26].

For immobilisation of enzymes via adsorption, the enzyme and the carrier should have large lipophilic surface areas. When an enzyme is immobilised on a hydrophobic carrier the force that binds them together cannot solely be contributed by van der Waal's forces

Materials Research Foundations **44** (2019) 51-72   doi: http://dx.doi.org/10.21741/9781644900079-3

which are weak in nature. It is rather the changes in entropy that leads to the efficient immobilisation of the enzyme. When an enzyme molecule interacts with a carrier, it displaces a large amount from itself and the carrier. The entropy gain resulting from such a reaction stabilized the immobilised particle and this interaction is called hydrophobic interaction. Lipase is a classic example that can be immobilised on a support via adsorption. A few examples of lipophilic carriers include EP-100 polypropylene, Accurel MP1004 propylene and octyl-silica. Lipases can be extracted, purified and immobilised from the crude enzyme extract by the addition of hydrophobic support media in a single step since more often than less lipase is the only lipophilic component [26]. Carriers such as octyl silica, Amberlite XAD 4, Accurel MP1000 and Lewatit VP OC 1600 etc. are hydrophobic in nature and are not readily dispensable in the enzyme suspension. For such carriers, it is best to pre-wet them with ethanol prior to immobilisation [27].

Hydrophilic amino acids on the surface of the enzyme molecule or glycosylation can result in increased hydrophilicity of the protein. This factor is taken into an advantage when immobilising enzymes onto hydrophilic supports via hydrogen bonding. Some of the most popular hydrophilic supports include avicel, cellulose, lignin, clay, silica, celite etc. Avicel, also known as calcium alginate, is a widely known simple and inexpensive support media that are used for enzyme immobilisation. This technology furthermore facilitates easier recovery of the immobilised enzyme (calcium alginate beads) from the reaction mixture. However, a problem faced with this material is the leakage of enzymes in an aqueous medium. This can be eliminated by cross-linking prior to immobilisation [28].

3.2   Cross-linking

Cross-linking of enzymes is an innovative method to immobilise enzymes without the help of a carrier material. Here, every enzyme molecule acts as a carrier for each other. Enzyme cross-linking can be achieved by first generating enzyme crystals by spray drying or in aqueous solutions. Cross-linked enzyme aggregates (CLEAs) are prepared by the addition of precipitants such as acetone, ethanol or 1, 2-dimethoxyethane and then using a crosslinker. Glutaraldehyde is a commonly used an agent to form intramolecular and intermolecular cross-links between proteins. Enzymes can, in some cases, be rendered inactive due to cross-linking. The efficiency of the cross-linking depends upon the amine content of the enzyme. Low amine content would mean ineffective cross-linking. However, studies have revealed that the inclusion of bovine serum albumin during the cross-linking reaction may reduce the extent of enzyme inactivation. Cross linked enzyme aggregates eliminate the disadvantages associated with carrier materials [28].

CLEA lipase produced from cocoa pod husk and has been to retain its activity even after six cycles. Furthermore, CLEAs has been reported to possess very high loading rate with little loss in enzyme activity. This technique can be developed into a protein purification mechanism as the contaminant proteins have harsher precipitation conditions compared to the target protein. However, high levels of protein purification cannot be expected. CLEAs tend to have several disadvantages compared to other immobilisation strategies. They are not mechanically resistant and may not be deemed fit for industrial applications. In cases of reactor media with high viscosity, the revival of the immobilised enzymes may be very difficult. These features of CLEAs make them an unlikely candidate for large-scale production processes [29]. Figure 1 represents how free enzymes are transformed into aggregates with the help of glutaraldehyde.

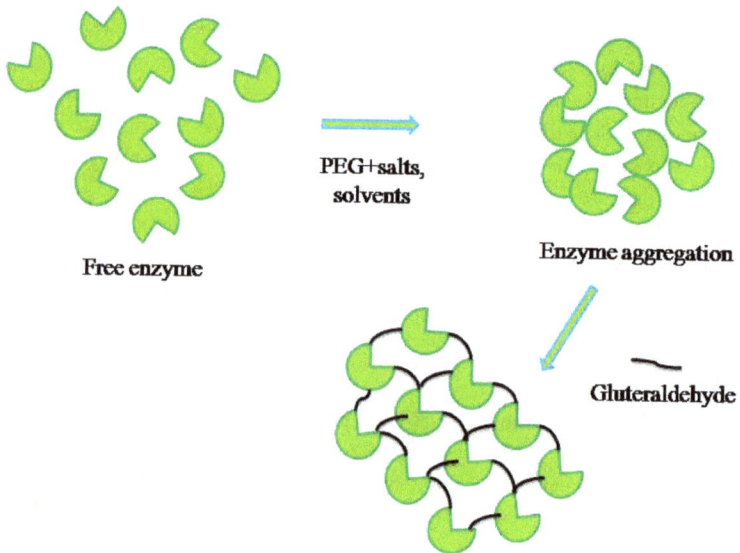

Figure 1. Cross linking enzyme aggregates formed employing gluteraldehyde as a linker

3.3   Covalent attachment

The covalent binding method basically fixes the enzyme molecules onto the substrate on a permanent basis through strong covalent bonds. This solves the problem of leaching of the enzyme found in adsorption-based biocatalysts when suspended in aqueous media. Thus, covalently immobilised enzymes can be used to great effect in reaction mixtures

Materials Research Forum LLC

doi: http://dx.doi.org/10.21741/9781644900079-3

that are aqueous in nature. Also, sometimes, covalent attachment requires the enzyme molecules to be bound to the carrier at multiple points. This stabilises the molecule in an environment where a native enzyme undergoes denaturation. In other words, multiple point covalent attachment of an enzyme molecule to the carrier increases its confirmation stability. Any hydrophobic or hydrophilic support can be used for covalent bond-based immobilisation technique. Amino acids with nucleophilic side chains found on the surface of the enzyme can bind to an epoxide or an aldehyde. Epoxy-activated supports are ideal for creating biocatalysts following covalent attachment method since they are stable during storage as well as in neutral aqueous media. Epoxy groups can attach to the support using short spacer arms which can then react with multiple points on the enzyme via different nucleophilic groups (amino, thiol, phenolic, imidazole etc.). The immobilisation process can then be easily be stopped by the addition of mercaptoethanol, ethanolamine or glycine, which attach to the epoxy groups [30].

Covalent immobilisation can be conducted in a site-directed manner to obtain biocatalysts with enhanced stability and reactivity. PRECISE (Protein Residue-Explicit Covalent Immobilisation for Stability Enhancement) system is a model that uses non-canonical amino acids and Huisgen 1, 3-dipolar cycloaddition 'click' reaction to enable site-directed enzyme immobilisation at residues that are rationally chosen through-out the enzyme. This system allows the user to immobilise the enzyme at proximate and distant locations from the active site to study the effect on activity and stability under harsh conditions. In a study evaluating the PRECISE system involving three variants of lysozyme, researchers were able to retain 70 to 85% of the activity of normal lysozyme. The problems posed by steric effect and steric hindrance were also nullified by this process which strongly suggested that immobilisation orientation has a pronounced influence over enzyme activity. Furthermore, subjecting the immobilised lysozyme variants to freeze-thaw cycles in the absence of a cryoprotectant with temperatures ranging from -80°C to 25°C did affect its activity to a certain extent. However, when the native enzymes only retained 20-30% of their activity, their immobilised counterparts were able to exhibit almost 70-80% retention of their activity [31].

3.4 Immobilisation via affinity tags

Enzymes, being proteins bear an inherent surface charge that is dependent upon their isoelectric points and the ambient pH values [32]. Surface charge and charge distribution of a protein can be determined by the use of various modelling software [33]. Based on the charge distribution and surface charge, enzymes can be attached to an ion exchanger via ionic and strongly polar interactions. Depending upon the charge of the enzyme, the ion exchanger may be positively charged or negatively charged. Not all enzymes possess a surface charge sufficient to strongly hold them to a support system. In cases like these, mutants have been constructed with an affinity tag on either the N terminal of the C terminal of the protein to aid purification and immobilisation [34]. His-tag, which is a common term with protein purification, is widely used for enzyme immobilisation. Histidine contains an imidazole ring which can act as a ligand for metal ions such as Cu^{2+}, Co^{2+} and Ni^{2+} to bind to the enzymes. This requires the support material to have metal chelates on their surface. Metal chelate epoxy surfaces are special kind of multifunctional epoxy supports used for the purification of enzymes with a poly His-tag attached to it. These support not only contain Ni^{2+} ions, but also Cu, Zn, Co, Fe^{2+} and Fe^{3+} ions which can interact with the His tag. Additionally, the dense layer of epoxy can react with the enzyme leading to enhanced binding strength [35]. The advantage of His tag is that it has little influence on the activity of the immobilised enzyme. This technique enables one-step purification of the enzyme and can be used for the design of microchannel reactors [36].

The problem with his-tag enzyme purification method is the requirement of heavy metals and hence it is not environmentally friendly. An alternative method suggests the use of aldehyde tags. Using recombinant DNA techniques, aldehyde tags can be introduced at the end of a protein molecule using the 6 amino acid consensus sequence which is recognised by the formyl glycine generating enzyme. Lipase enzyme was immobilised and purified by coupling the aldehyde from the lipase polypeptide with the support through Schiff's base reaction. The support material used here was amino-functionalized mesocellular siliceous foam (MCS) so that the active site could be protected from exposure to the cross linker [37].

4.   Entrapment

Entrapment can be described as the engulfing of the enzyme inside a matrix such as gels or fibres. This method does not involve chemically bonding of the enzyme to the carrier material. The three-dimensional structure is therefore preserved upon the immobilisation process and the optimum reaction conditions such as pH, temperature, enzyme concentration and incubation time are left unaltered. A typical example of entrapment method is the use of poly anionic or cationic polymers to form gels by the addition of counter ions [38]. Polyurethane, silicone and latex material have been reported for various enzyme immobilisation studies. Enzymes such as β-glucosidase were immobilised in latex and silicone polymers through entrapment method with improvement in stability and reusability [39]. Agar agar was tested as a carrier material for endo-β-1,4-xylanase with an increase in optimum temperature, reaction time, and extended stability at different temperature ranges [40].

Materials Research Forum LLC

doi: http://dx.doi.org/10.21741/9781644900079-3

Several variations of entrapment have been described by which enzymes can be successfully entrapped in a solid support without any loss in optimum activity. Dry entrapment is one such technique that was developed for water-free enzyme preparations. Standard polymeric glue such as polyester or epoxy-based resin has been used for this purpose with water dry enzymes. The supports are coated with the enzyme/polymer mix for polymerisation and then tested with normal enzyme activity assay techniques. Enzymes to be immobilised can be used in the freely diffusing form, adsorbed on octyl sepharose beads or as cross-linked enzyme aggregates (CLEAs). This method can be used for lipase derived from *Thermomyces lanuginosus* and L-threonine aldolase from *Aspergillus gossypii*. Studies have confirmed that octyl sepharose and octyle agarose are perfect support materials for this method. The activity of immobilised *Thermomyces lanuginosus* lipase towards a substrate depends upon the length of the fatty acid chains. Immobilised lipases tend to show higher activity towards short chain fatty acids in comparison to long chains due to side effects and diffusion limitation. Enzyme recovery has been a major concern with immobilising enzymes using dry entrapment [41].

5. Microencapsulation

Encapsulation is the best way to protect the enzyme from any negative effects contributed by the reactor environment. Enzyme encapsulation has been reported using several materials such as calcium alginate and polyvinyl alcohol, but the sol-gel technique has emerged as the popular one. A diversity of studies has been conducted with enzyme encapsulation based on sol-gels. Sol-gels are silica materials that are porous in nature and are benign to enzymes. These gels do not involve in any chemical bond formation with the enzyme, prevent enzyme leakage, the tunable porosity, mechanical stability and negligible swelling behaviour make silica sol-gels as the preferred material for enzyme encapsulation [42].

Sol-gels are prepared using tetramethyl silane (TMOS) as the precursor material. TMOS is hydrolysed using acid followed by condensation to form the sol. The sol is a mixture of partially hydrolysed and partially condensed monomer molecules. The gel is formed when the condensation continues. When all the pores in the gel are filled with water and alcohol, it forms the aqua gel. The water in the aqua gel can be evaporated by capillary action and when this happens a part of it collapses to form the xerogel. On the other hand, if the water is replaced by acetone and then supercritical carbon dioxide, the structure of the aqua gel is maintained on evaporation of the CO_2 and the brittle aerogel is formed [43]. The addition of alkyltrialkoxysilanes to the synthesis mixture can lead to the formation of hydrophobic surfaces on sol-gels.

This technique has been tested as an enzyme immobilisation measure for many enzymes. Hydrozynitrile lyase (HNL) is an important industrial enzyme due to its ability to synthesise enantiopure cyanohydrins from aldehydes/ketones and HCN. An attempt was made to immobilise HNL from *Hevea brasiliensis.* Ordinary procedures for the production of sol-gels resulted in the enzyme getting denatured as it was sensitive to the methanol that is expelled during the formation of sol. A new technique was therefore devised where the alkoxy silanes were almost 100% hydrolysed mediated by acid. This resulted in the removal of methanol by evaporation. The enzyme dissolved in a buffer with pH 6.5 was mixed in the precursor sol. The rise in pH accelerated the condensation reaction and the enzymes were stabilised. As soon as the gel was formed it was submerged in the buffer with pH 6.5 to remove any excess methanol that may have formed by the hydrolysis of the methoxy groups. This procedure was successful in retaining 65% activity of the native enzyme in standard aqueous activity tests. The loss in activity can be attributed to methanol deactivation and diffusion limitations [44].

6.   One step immobilisation and purification

Enzymes can be immobilised and purified in a single step and techniques that aid this process are becoming more and more popular. In general terms, a single interaction between the enzyme and support material is all it takes to stabilize the enzyme and at other times several interactions may be necessary for the protein molecule to remain attached to the support. But as it is obvious, the focus of enzyme immobilisation should be multipoint or multi subunit attachment to improve enzyme stability [45]. Enzyme immobilisation and purification via one step can be achieved by following three different strategies: immobilisation via one point, employing custom-made supports that are specific to the target protein based on certain structural features by the formulation of heterofunctional supports to immobilise a specific enzyme via multipoint attachment and, the application of site direct mutagenesis in the effort to introduce specific domains in the target protein molecule that show affinity to the heterofunctional supports. The immobilisation of lipases on hydrophobic supports by means of interfacial activation is a special case.

An example of a single point immobilisation is the use of antigen-antibody interaction. This is a selective process with high levels of sensitivity because only the target protein becomes immobilised. Monoclonal or polyclonal antibodies are used for this purpose. Two factors that govern the success of this procedure (i) the immobilisation of the target protein on the antibody and (ii) the prevention of any undesired adsorption. The use of monoclonal antibodies gives the user the flexibility to decide the orientation of the enzyme with respect to the support surface. This is an important feature since it permits the user to choose the orientation where the active site of the enzyme is fully exposed [46]. This may also help in safeguarding the enzyme if the site of attachment is a fragile region which can be inactivated by inhibitors. The immobilisation yield following this procedure is 100% and almost pure immobilised enzyme is achieved [47]. In some cases, the antibody is not specific to the target protein but to a certain domain that has been introduced by site-directed mutagenesis. Using this step, the correct orientation of the target protein can be guaranteed. Also, any protein with the required domain can be immobilised on this support. In this case, as well, the immobilisation is almost 100% with the achievement of total purity of the enzyme but the stabilisation is not that significant since there is no interaction between the enzyme and the support material [48].

The second technique is by the introduction of different domains that induce affinity between the protein molecule and the support. These domains are peptides that are very small e.g. cellulose binding domain and other artificial peptides that are proteinaceous in nature like his-tag. Affinity immobilisation using tags have already been covered in an earlier section of this chapter. In the case of lipases, it is possible to take advantage of specific catalytic mechanisms. Interfacial activation is a mechanism used by lipases to act on the surface of oil drops. Lipases have very large hydrophobic active centres and hydrophilic exteriors. In an aqueous environment, the hydrophobic active centres are protected by a lid (a polypeptide chain). Meanwhile, the hydrophilic exterior is exposed to the aqueous environment. In this conformation, the lipase is essentially closed. However, there is yet another form called the open form where the hydrophobic catalytic centre is exposed; both forms shift between each other and are in equilibrium. So, in the presence of an oil drop the lid is displaced exposing the hydrophobic catalytic core and the equilibrium taking up an open confirmation. This property of lipases where they can adsorb on to hydrophobic surfaces is called interfacial activation [49-51].

Interfacial activation can be induced by the introduction of any form on a hydrophobic substance such as an oil drop or a hydrophobic support or even a hydrophobic protein for that matter. All these materials have extensively been used for the purification of lipases. The immobilisation of lipases on a hydrophobic support at low ionic strength permits the stabilisation of the open form on the enzyme in one single step and therefore is a much-applied step for purification [52, 53]. This process works only in low ionic strengths. High ionic strength during immobilisation/purification favours the lipase to be in closed form and favours the adsorption of other proteins on to the substrate. Conventional adsorption process of lipase is not as efficient as interfacial activation [54]. Based on this criterion, lipases behave in different ways during adsorption on a hydrophobic support. Some lipases require highly hydrophobic supports while others can only be adsorbed on mild ones. The nature of the group on the support can also be a determining factor for lipase purification. Usually, octyl agarose and sepabeads decaoctyl are the supports that can contribute to lipase 1000-fold stability [55]. This protocol is very simple and can easily be used on an industrial level. One of the most popular commercialised immobilised lipase preparations, Novozym 435, is achieved by using this technique.

In some cases, immobilisation of the protein by just one link between the enzyme and the support is desired to improve enzyme stability. In situations like these, heterofunctional groups are desired with the tag interacting with one domain of the enzyme while other groups get involved in the formation of multipoint covalent bonds [56]. Heterofunctional supports are matrices with several functional groups on their surface capable of physical and chemical interactions intended to interact with the protein of interest. Some of the groups present adsorbs the target protein while the rest of the functionalities contribute to making the protein-support interaction irreversible. This strategy enables the purification of the protein by selective adsorption. This method is applicable to large proteins with multimeric subunits. Epoxides and glyoxyl groups are the chemically reactive groups used in this mode of immobilisation/purification. Glyoxyl groups cannot immobilise proteins in neutral pH [57-59]. After adsorption, the enzyme may interact with the support groups to form covalent bonds contributing to an increase in stability. Increasing the pH can result in an increase in the reactivity of nucleophilic groups on the surface of the enzyme on to the support [60].

Conclusions and future trends

To summarise, enzyme immobilisation is of utmost importance when it comes to the economics of production processes and as far as the environment is concerned. A better understanding of the chemical reactions involved in enzyme-support interactions provides us with a better idea about the choice of carrier material and mode of immobilisation. Magnetic nanoparticles and carbon nanotubes are becoming increasingly interesting supports for proteins [61]. Conventional support materials can be coupled with magnetic nanoparticles to facilitate easier separation of the immobilysate [62]. Technologies have emerged that enable linking of the support groups to the protein subunits at specific sites. Multipoint linkages between enzymes and support impart better stability to the finished product. Heterofunctional supports to facilitate the formation of multiple bonds of various natures between the protein and the carrier without skewing the enzyme conformation. Therefore, the introduction of heterofunctional supports for industrial production of immobilised enzymes will be an interesting prospect. However, the use of emerging methods in the commercial production of immobilised enzymes is still at a very early stage. The most promising techniques need to be scaled up to determine whether they can replicate the same results obtained in laboratory and pilot scale.

Acknowledgement

Authors would like to acknowledge the funding from Dublin Institute of Technology (DIT) under the Fiosraigh Scholarship programme, 2014.

References

[1]   E. García-Urdiales, I. Alfonso, V. Gotor, Update 1 of: enantioselective enzymatic desymmetrizations in organic synthesis, Chem. Rev. 11 (2011) PR110–PR180. https://doi.org/10.1021/cr100330u

[2]   C.J. Malemud, Matrix metalloproteinases (MMPs) in health and disease: an overview, Frontiers In Bioscience: A Journal And Virtual Library. 11 (2005) 1696-1701. https://doi.org/10.2741/1915

[3]   A. Wolfgang, Enzyme in industry: Production and Applications, Wiley-VCH, Weinheim, 2007.

[4]   G. Haki, S. Rakshit, Developments in industrially important thermostable enzymes: a review, Bioresource Technology 89 (2003) 17-34. https://doi.org/10.1016/S0960-8524(03)00033-6

[5]   I. Alkorta, C. Garbisu, M.J. Llama, J.L. Serra, Industrial applications of pectic enzymes: a review, Process Biochem. 33 (1998) 21-28. https://doi.org/10.1016/S0032-9592(97)00046-0

[6]   H. Neurath, K.A. Walsh, Role of proteolytic enzymes in biological regulation (a review), Proceedings of the National Academy of Sciences 73 (1976) 3825-3832. https://doi.org/10.1073/pnas.73.11.3825

[7]   J.R. Cherry, A.L. Fidantsef, Directed evolution of industrial enzymes: an update, Current Opinion in Biotechnology 14 (2003) 438-443. https://doi.org/10.1016/S0958-1669(03)00099-5

[8]   B. Joseph, P.W. Ramteke, G. Thomas, Cold active microbial lipases: some hot issues and recent developments, Biotechnology Advances 26 (2008) 457-470. https://doi.org/10.1016/j.biotechadv.2008.05.003

[9]   J. Lalonde, A. Margolin, Immobilization of enzymes, Enzyme Catalysis in Organic Synthesis: A Comprehensive Handbook, Second Edition (2008) 163-184.

[10] N. Burgoyne, R. Jackson, Predicting Protein Function from Surface Properties, in: D. Rigden (Ed.) From Protein Structure to Function with Bioinformatics, Springer Netherlands 2009, pp. 167-186. https://doi.org/10.1007/978-1-4020-9058-5_7

[11] L. Cao, Carrier-bound immobilized enzymes: principles, application and design, John Wiley & Sons 2006.

[12] U. Hanefeld, L. Gardossi, E. Magner, Understanding enzyme immobilisation, Chemical Society Reviews 38 (2009) 453-468. https://doi.org/10.1039/B711564B

[13] R.A. Sheldon, S. van Pelt, Enzyme immobilisation in biocatalysis: why, what and how, Chemical Society Reviews 42 (2013) 6223-6235. https://doi.org/10.1039/C3CS60075K

[14] R.Y. Cabrera-Padilla, M.C. Lisboa, A.T. Fricks, E. Franceschi, A.S. Lima, D.P. Silva, C.M. Soares, Immobilization of Candida rugosa lipase on poly (3-hydroxybutyrate-co-hydroxyvalerate): A new eco-friendly support, Journal of Industrial Microbiology & Biotechnology 39 (2012) 289-298. https://doi.org/10.1007/s10295-011-1027-3

[15] G.-W. Xing, X.-W. Li, G.-L. Tian, Y.-H. Ye, Enzymatic peptide synthesis in organic solvent with different zeolites as immobilization matrixes, Tetrahedron 56 (2000) 3517-3522. https://doi.org/10.1016/S0040-4020(00)00261-1

[16] J.F. Díaz, K.J. Balkus, Enzyme immobilization in MCM-41 molecular sieve, Journal of Molecular Catalysis B: Enzymatic 2 (1996) 115-126. https://doi.org/10.1016/S1381-1177(96)00017-3

[17] F. Yagiz, D. Kazan, A.N. Akin, Biodiesel production from waste oils by using lipase immobilized on hydrotalcite and zeolites, Chemical Engineering Journal 134 (2007) 262-267. https://doi.org/10.1016/j.cej.2007.03.041

[18] N. Carlsson, H. Gustafsson, C. Thörn, L. Olsson, K. Holmberg, B. Åkerman, Enzymes immobilized in mesoporous silica: A physical–chemical perspective, Advances in Colloid and Interface Science 205 (2014) 339-360. https://doi.org/10.1016/j.cis.2013.08.010

[19] B. Kalska-Szostko, M. Rogowska, A. Dubis, K. Szymański, Enzymes immobilization on Fe3O4–gold nanoparticles, Applied Surface Science 258 (2012) 2783-2787. https://doi.org/10.1016/j.apsusc.2011.10.132

Enzymatic Fuel Cells
Materials Research Foundations **44** (2019) 51-72

Materials Research Forum LLC
doi: http://dx.doi.org/10.21741/9781644900079-3

[20]   E. Biró, Á.S. Németh, C. Sisak, T. Feczkó, J. Gyenis, Preparation of chitosan particles suitable for enzyme immobilization, Journal of Biochemical and Biophysical Methods 70 (2008) 1240-1246. https://doi.org/10.1016/j.jprot.2007.11.005

[21]   E.W. Nery, L.T. Kubota, Evaluation of enzyme immobilization methods for paper-based devices—A glucose oxidase study, Journal of Pharmaceutical and Biomedical Analysis 117 (2016) 551-559. https://doi.org/10.1016/j.jpba.2015.08.041

[22]   P. Hanachi, F. Jafary, F. Jafary, S. Motamedi, Immobilization of the alkaline phosphatase on collagen surface via cross-linking method, Iranian Journal of Biotechnology 13 (2015) 32-38. https://doi.org/10.15171/ijb.1203

[23]   T. Coradin, J. Livage, Mesoporous alginate/silica biocomposites for enzyme immobilisation, Comptes Rendus Chimie 6 (2003) 147-152. https://doi.org/10.1016/S1631-0748(03)00006-7

[24]   T. Jesionowski, J. Zdarta, B. Krajewska, Enzyme immobilization by adsorption: A review, Adsorption 20 (2014) 801-821. https://doi.org/10.1007/s10450-014-9623-y

[25]   H.H.P. Yiu, P.A. Wright, N.P. Botting, Enzyme immobilisation using SBA-15 mesoporous molecular sieves with functionalised surfaces, Journal of Molecular Catalysis B: Enzymatic 15 (2001) 81-92. https://doi.org/10.1016/S1381-1177(01)00011-X

[26]   M. Persson, E. Wehtje, P. Adlercreutz, Immobilisation of lipases by adsorption and deposition: high protein loading gives lower water activity optimum, Biotechnology letters 22 (2000) 1571-1575. https://doi.org/10.1023/A:1005689002238

[27]   T. Zhao, D.S. No, B.H. Kim, H.S. Garcia, Y. Kim, I.-H. Kim, Immobilized phospholipase A1-catalyzed modification of phosphatidylcholine with n−3 polyunsaturated fatty acid, Food Chemistry 157 (2014) 132-140. https://doi.org/10.1016/j.foodchem.2014.02.024

[28]   C.-T. Tsai, A.S. Meyer, Enzymatic cellulose hydrolysis: Enzyme reusability and visualization of β-Glucosidase immobilized in calcium alginate, Molecules 19 (2014) 19390-19406. https://doi.org/10.3390/molecules191219390

[29]   S. Khanahmadi, F. Yusof, A. Amid, S.S. Mahmod, M.K. Mahat, Optimized
preparation and characterization of CLEA-lipase from cocoa pod husk, Journal of
Biotechnology 202 (2015) 153-161. https://doi.org/10.1016/j.jbiotec.2014.11.015

[30]   C. Mateo, O. Abian, R. Fernandez–Lafuente, J.M. Guisan, Increase in
conformational stability of enzymes immobilized on epoxy-activated supports by
favoring additional multipoint covalent attachment, Enzyme and Microbial
Technology 26 (2000) 509-515. https://doi.org/10.1016/S0141-0229(99)00188-X

[31]   J.C.Y. Wu, C.H. Hutchings, M.J. Lindsay, C.J. Werner, B.C. Bundy, Enhanced
enzyme stability through site-directed covalent immobilization, Journal of
Biotechnology 193 (2015) 83-90. https://doi.org/10.1016/j.jbiotec.2014.10.039

[32]   R.M. Kramer, V.R. Shende, N. Motl, C.N. Pace, J.M. Scholtz, Toward a molecular
understanding of protein solubility: increased negative surface charge correlates
with increased solubility, Biophysical Journal 102 (2012) 1907-1915.
https://doi.org/10.1016/j.bpj.2012.01.060

[33]   Y.R. Gokarn, R.M. Fesinmeyer, A. Saluja, V. Razinkov, S.F. Chase, T.M. Laue,
D.N. Brems, Effective charge measurements reveal selective and preferential
accumulation of anions, but not cations, at the protein surface in dilute salt
solutions, Protein Science 20 (2011) 580-587. https://doi.org/10.1002/pro.591

[34]   C.M. Halliwell, E. Simon, C.-S. Toh, A.E.G. Cass, P.N. Bartlett, The design of
dehydrogenase enzymes for use in a biofuel cell: the role of genetically introduced
peptide tags in enzyme immobilization on electrodes, Bioelectrochemistry 55
(2002) 21-23. https://doi.org/10.1016/S1567-5394(01)00172-4

[35]   N. Bortone, M. Fidaleo, Immobilization of the recombinant (His)6-tagged l-
arabinose isomerase from Thermotoga maritima on epoxy and cupper-chelate
epoxy supports, Food and Bioproducts Processing 95 (2015) 155-162.
https://doi.org/10.1016/j.fbp.2015.05.002

[36]   M. Miyazaki, J. Kaneno, S. Yamaori, T. Honda, M.P. P. Briones, M. Uehara, K.
Arima, K. Kanno, K. Yamashita, Y. Yamaguchi, H. Nakamura, H. Yonezawa, M.
Fujii, H. Maeda, efficient immobilization of enzymes on microchannel surface
through his-tag and application for microreactor, Protein and Peptide Letters 12
(2005) 207-210. https://doi.org/10.2174/0929866053005854

[37] A. Wang, F. Du, F. Wang, Y. Shen, W. Gao, P. Zhang, Convenient one-step purification and immobilization of lipase using a genetically encoded aldehyde tag, Biochemical Engineering Journal 73 (2013) 86-92. https://doi.org/10.1016/j.bej.2013.02.003

[38] H. Kamal, E.L.S.A. Hegazy, H.M. Sharada, S.A. Abd Elhalim, S. Lotfy, R.D. Mohamed, Immobilization of glucose isomerase onto radiation synthesized P(AA-co-AMPS) hydrogel and its application, Journal of Radiation Research and Applied Sciences 7 (2014) 154-162. https://doi.org/10.1016/j.jrras.2014.02.001

[39] M.R. Javed, A. Buthe, M.H. Rashid, P. Wang, Cost-efficient entrapment of β-glucosidase in nanoscale latex and silicone polymeric thin films for use as stable biocatalysts, Bioproducts and Biosystems Engineering 190 (2016) 1078-1085. https://doi.org/10.1016/j.foodchem.2015.06.040

[40] Z. Bibi, F. Shahid, S.A. Ul Qader, A. Aman, Agar–agar entrapment increases the stability of endo-β-1,4-xylanase for repeated biodegradation of xylan, International Journal of Biological Macromolecules 75 (2015) 121-127. https://doi.org/10.1016/j.ijbiomac.2014.12.051

[41] S. Barig, A. Funke, A. Merseburg, K. Schnitzlein, K.P. Stahmann, Dry entrapment of enzymes by epoxy or polyester resins hardened on different solid supports, Enzyme and Microbial Technology 60 (2014) 47-55. https://doi.org/10.1016/j.enzmictec.2014.03.013

[42] I. Mazurenko, W. Ghach, G.-W. Kohring, C. Despas, A. Walcarius, M. Etienne, Immobilization of membrane-bounded (S)-mandelate dehydrogenase in sol–gel matrix for electroenzymatic synthesis, Bioelectrochemistry 104 (2015) 65-70. https://doi.org/10.1016/j.bioelechem.2015.03.004

[43] E.I. Goksu, M.I. Hoopes, B.A. Nellis, C. Xing, R. Faller, C.W. Frank, S.H. Risbud, J.H. Satcher Jr, M.L. Longo, Silica xerogel/aerogel-supported lipid bilayers: Consequences of surface corrugation, Biochimica et Biophysica Acta (BBA) - Biomembranes 1798 (2010) 719-729. https://doi.org/10.1016/j.bbamem.2009.09.007

[44] L. Veum, U. Hanefeld, A. Pierre, The first encapsulation of hydroxynitrile lyase from Hevea brasiliensis in a sol–gel matrix, Tetrahedron 60 (2004) 10419-10425. https://doi.org/10.1016/j.tet.2004.06.135

Materials Research Forum LLC

doi: http://dx.doi.org/10.21741/9781644900079-3

[45]   C. Garcia-Galan, Á. Berenguer-Murcia, R. Fernandez-Lafuente, R.C. Rodrigues, Potential of different enzyme immobilization strategies to improve enzyme performance, Advanced Synthesis and Catalysis 353 (2011) 2885-2904. https://doi.org/10.1002/adsc.201100534

[46]   K. Hernandez, R. Fernandez-Lafuente, Control of protein immobilization: Coupling immobilization and site-directed mutagenesis to improve biocatalyst or biosensor performance, Enzyme and Microbial Technology 48 (2011) 107-122. https://doi.org/10.1016/j.enzmictec.2010.10.003

[47]   A. Fatima, Q. Husain, Polyclonal antibodies mediated immobilization of a peroxidase from ammonium sulphate fractionated bitter gourd (Momordica charantia) proteins, Biomolecular Engineering 24 (2007) 223-230. https://doi.org/10.1016/j.bioeng.2006.10.002

[48]   J. Wang, D. Bhattacharyya, L.G. Bachas, Orientation specific immobilization of organophosphorus hydrolase on magnetic particles through gene fusion, Biomacromolecules 2 (2001) 700-705. https://doi.org/10.1021/bm015517x

[49]   H. van Tilbeurgh, M.-P. Egloff, C. Martinez, N. Rugani, R. Verger, C. Cambillau, Interfacial activation of the lipase–procolipase complex by mixed micelles revealed by X-ray crystallography, Nature 362 (1993) 814–820. https://doi.org/10.1038/362814a0

[50]   A. Brzozowski, U. Derewenda, Z. Derewenda, G. Dodson, D. Lawson, J. Turkenburg, F. Bjorkling, B. Huge-Jensen, S. Patkar, L. Thim, A model for interfacial activation in lipases from the structure of a fungal lipase-inhibitor complex, Nature 351 (1991) 491-494. https://doi.org/10.1038/351491a0

[51]   Z.S. Derewenda, U. Derewenda, G.G. Dodson, The crystal and molecular structure of the Rhizomucor miehei triacylglyceride lipase at 1.9 Å resolution, Journal of Molecular Biology 227 (1992) 818-839. https://doi.org/10.1016/0022-2836(92)90225-9

[52]   G. Fernández-Lorente, J.M. Palomo, Z. Cabrera, J.M. Guisán, R. Fernández-Lafuente, Specificity enhancement towards hydrophobic substrates by immobilization of lipases by interfacial activation on hydrophobic supports, Enzyme and Microbial Technology 41 (2007) 565-569. https://doi.org/10.1016/j.enzmictec.2007.05.004

[53] G. Fernandez-Lorente, Z. Cabrera, C. Godoy, R. Fernandez-Lafuente, J.M. Palomo, J.M. Guisan, Interfacially activated lipases against hydrophobic supports: Effect of the support nature on the biocatalytic properties, Process Biochemistry 43 (2008) 1061-1067. https://doi.org/10.1016/j.procbio.2008.05.009

[54] E.A. Manoel, J.C. dos Santos, D.M. Freire, N. Rueda, R. Fernandez-Lafuente, Immobilization of lipases on hydrophobic supports involves the open form of the enzyme, Enzyme and Microbial Technology 71 (2015) 53-57. https://doi.org/10.1016/j.enzmictec.2015.02.001

[55] J.M. Palomo, G. Muñoz, G. Fernández-Lorente, C. Mateo, R. Fernández-Lafuente, J.M. Guisán, Interfacial adsorption of lipases on very hydrophobic support (octadecyl–Sepabeads): immobilization, hyperactivation and stabilization of the open form of lipases, Journal of Molecular Catalysis B: Enzymatic 19 (2002) 279-286. https://doi.org/10.1016/S1381-1177(02)00178-9

[56] J.M. Bolivar, J. Rocha-Martin, C. Godoy, R.C. Rodrigues, J.M. Guisan, Complete reactivation of immobilized derivatives of a trimeric glutamate dehydrogenase from Thermus thermophillus, Process Biochemistry 45 (2010) 107-113. https://doi.org/10.1016/j.procbio.2009.08.014

[57] O. Barbosa, R. Torres, C. Ortiz, A.n. Berenguer-Murcia, R.C. Rodrigues, R. Fernandez-Lafuente, Heterofunctional supports in enzyme immobilization: From traditional immobilization protocols to opportunities in tuning enzyme properties, Biomacromolecules 14 (2013) 2433-2462. https://doi.org/10.1021/bm400762h

[58] C. Mateo, J.M. Palomo, M. Fuentes, L. Betancor, V. Grazu, F. López-Gallego, B.C. Pessela, A. Hidalgo, G. Fernández-Lorente, R. Fernández-Lafuente, Glyoxyl agarose: A fully inert and hydrophilic support for immobilization and high stabilization of proteins, Enzyme and Microbial Technology 39 (2006) 274-280. https://doi.org/10.1016/j.enzmictec.2005.10.014

[59] C. Mateo, O. Abian, M. Bernedo, E. Cuenca, M. Fuentes, G. Fernandez-Lorente, J.M. Palomo, V. Grazu, B.C. Pessela, C. Giacomini, Some special features of glyoxyl supports to immobilize proteins, Enzyme and Microbial Technology 37 (2005) 456-462. https://doi.org/10.1016/j.enzmictec.2005.03.020

[60]   J. Pedroche, M. del Mar Yust, C. Mateo, R. Fernández-Lafuente, J. Girón-Calle, M. Alaiz, J. Vioque, J.M. Guisán, F. Millán, Effect of the support and experimental conditions in the intensity of the multipoint covalent attachment of proteins on glyoxyl-agarose supports: correlation between enzyme–support linkages and thermal stability, Enzyme and Microbial Technology 40 (2007) 1160-1166. https://doi.org/10.1016/j.enzmictec.2006.08.023

[61]   A.P.M. Tavares, C.G. Silva, G. Dražić, A.M.T. Silva, J.M. Loureiro, J.L. Faria, Laccase immobilization over multi-walled carbon nanotubes: Kinetic, thermodynamic and stability studies, Journal of Colloid and Interface Science 454 (2015) 52-60. https://doi.org/10.1016/j.jcis.2015.04.054

[62]   E.-J. Woo, H.-S. Kwon, C.-H. Lee, Preparation of nano-magnetite impregnated mesocellular foam composite with a Cu ligand for His-tagged enzyme immobilization, Chemical Engineering Journal 274 (2015) 1-8. https://doi.org/10.1016/j.cej.2015.03.123

Enzymatic Fuel Cells
Materials Research Foundations **44** (2019) 73-108

Materials Research Forum LLC
doi: http://dx.doi.org/10.21741/9781644900079-4

Chapter 4

Fuel Cell Electrochemistry

Fatma Aydin Unal[1,2], Hakan Burhan[1], Neslihan Karaman[1], Kubilay Arıkan[1], Bahar Simsek[1],
Burcu Akyıldız[1], Fatih Şen[1]*

[1] Sen Research Group, Biochemistry Department, Faculty of Arts and Science, Dumlupınar
University, Evliya Çelebi Campus, 43100 Kütahya, Turkey

[2] Metallurgical and Materials Engineering Department, Faculty of Engineering, Alanya Alaaddin
Keykubat University, 07450 Alanya/Antalya, Turkey

fatihsen1980@gmail.com

Abstract

Fuel cells can be defined as devices which convert chemical energy into electrical energy.
They have a potential capability for promising energy systems. Especially, as a clean and
renewable energy source, fuel cells have very significant importance. There are many
types of fuel cells such as alcohol, hydrogen, biofuel cells etc. This chapter examines the
general electrochemistry of fuel cells. Furthermore, the basis of a fuel cell, fuel cell
chemistry, substrates and potentials, electrochemistry, types of the electrode,
potentiostats have been examined in detail. Besides, the electrochemistry characterization
techniques, polarisation and power curves, microbial fuel cell and their reactions,
limitations of the electrochemical reactions are also described in detail.

Keywords

Electrochemistry, Fuel Cell Electrochemistry, Microbial Fuel Cell, Fuel Cell Reaction

Contents

Materials Research Forum LLC
doi: http://dx.doi.org/10.21741/9781644900079-4

1.   Introduction

Fuel cells can be described as devices which convert chemical energy into electrical energy. They have a potential capability for promising energy systems. Especially, as a clean and renewable energy source, fuel cells have very significant importance. There are many types of fuel cells such as alcohol fuel cells, solid oxide fuel cells, hydrogen fuel cells, biological fuel cells, etc. [1]. Biological fuel cells use biocatalysts for the conversion of chemical energy to electrical energy [2]. Biofuel cells are classified into two different categories. One of them is called a microbial fuel cell, and the other is called enzymatic fuel cell [3]. Fuel cells that use a specified enzyme are known as enzymatic fuel cells (EFCs) while systems that use whole microorganisms are known as microbial fuel cells (MFCs) [4–11].

Enzymatic Fuel Cells   Materials Research Forum LLC
Materials Research Foundations **44** (2019) 73-108   doi: http://dx.doi.org/10.21741/9781644900079-4

1.1   Basis of a Fuel Cell

Electrochemistry is a branch of chemistry which addresses the interrelation of chemical and electrical effects. Some of the electrochemical reactions are directly related to the fuel cells and the basic of the fuel cell can be explained by the following reaction [12];

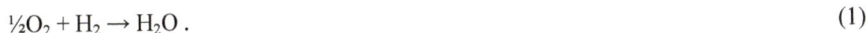

$$\tfrac{1}{2}O_2 + H_2 \rightarrow H_2O . \tag{1}$$

In this reaction, generally oxidation-reduction reaction mechanism occurs and this is the fundamental of fuel cells. For the oxidation-reduction reactions, some of the kinetic parameters such as entalphy, entropy, etc. can be defined as in the following reaction:

$$\Delta H = \Delta G + T. \Delta S \tag{2}$$

Another important parameter in electrochemical systems is the efficiency (η) and it can be calculated by the following equation;

$$\eta = \frac{\Delta G}{\Delta H} \tag{3}$$

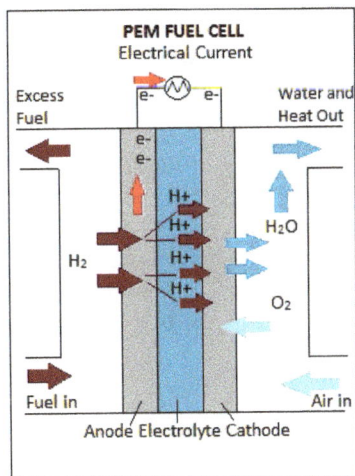

Figure 1. Working principle of hydrogen fuel cell with PEM technology.

Materials Research Forum LLC
doi: http://dx.doi.org/10.21741/9781644900079-4

Besides, in the fuel cells, hydrogen can act as a fuel in anodic part of the cells; oxygen is generally in cathodic part of the fuel cells as shown in following reactions and also in Figure 1.

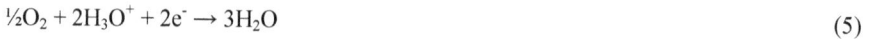

$$H_2 + 2H_2O \rightarrow 2H_3O^+ + 2e^- \tag{4}$$

$$\tfrac{1}{2}O_2 + 2H_3O^+ + 2e^- \rightarrow 3H_2O \tag{5}$$

The potential of these fuel cells can be calculated with the help of following Nernst's equation [12].

$$W = n.F.E = -\Delta G \tag{6}$$

1.2 Fuel Cell Chemistry

There are many types of fuel cells such as alcohol fuel cells, solid oxide fuel cells, hydrogen fuel cells and biofuel cells. Biofuel cells are also classified into two different categories. One of them is called a microbial fuel cell, and the other is called enzymatic fuel cell. Fuel cells that use a specified enzyme are known as enzymatic fuel cells (EFCs) while systems that use whole microorganisms are known as microbial fuel cells (MFCs). One of the most studied organisms in microbial fuel cells is *Geobacter* sulfurreducens which is the most promising one in terms of power density. The preferred substrate for this organism is acetate [12]. Half-cell reactions for these organisms are as follows [13–14]:

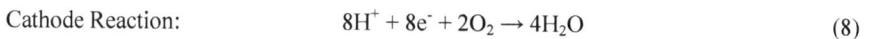

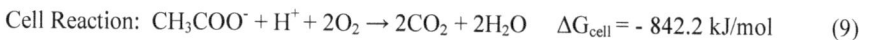

Anode Reaction: $$CH_3COO^- + H^+ + 2H_2O \rightarrow 2CO_2 + 8H^+ + 8\,e^- \tag{7}$$

Cathode Reaction: $$8H^+ + 8e^- + 2O_2 \rightarrow 4H_2O \tag{8}$$

Cell Reaction: $CH_3COO^- + H^+ + 2O_2 \rightarrow 2CO_2 + 2H_2O \quad \Delta G_{cell} = -842.2 \text{ kJ/mol} \tag{9}$

Using the Nernst equation, this Gibbs free energy value corresponds to the following theoretical cell voltage:

$$E_{cell} = \frac{-\Delta G_{cell}}{nF} = \frac{-(-842\,000\frac{J}{mol})}{(8\ mol\ e^-)(96485\frac{C}{mol\ e^-}}$$ (10)

$$E_{cell} = 1.09\ V.$$ (11)

This is the maximum possible voltage for this type of microbial fuel cells with an oxygen cathode system. However, in practice, the open circuit potential typically has a much lower value between 0.6 V and 0.8 V. This inconsistency is caused by loss of metabolism and energy induced by the bacteria, and these bacteria are inevitable if they gain any benefit without breathing in the anode. It is also assumed that this formula is a pure oxygen cathode instead of an air cathode, and activity for all species present (implying 1 M acetate, pH 0) [12]. Even corrected to reflect the expected acetate concentration and pH value, it is difficult to predict them accurately, since the intracellular acetate concentrations are more critical than the extracellular ones and difficult to measure [15–17].

1.3   Substrates and Potentials

In microbial fuel cells, very simple materials such as glucose, acetate, fructose, etc. can be used as substrates. These are anaerobically amplified to force the cells to use an electrode as the last electron receiver instead of dissolved oxygen. The maximum theoretical potential (voltage) depends upon the materials, its concentration, and the type of cathode [18]. The theoretical potential of the anode is calculated from the reference potentials of the reaction components or the Gibbs free reaction energy [19] as shown below:

$$E° = \frac{-\Delta G}{nF}$$ (11)

When combined with the Nernst equation and the reaction coefficient, this can be used to determine the theoretical potential in non-standard conditions [20]:

$$E = \frac{-\Delta G}{nF} - \frac{RT}{nF}\ln(\Pi)$$ (12)

Materials Research Forum LLC
doi: http://dx.doi.org/10.21741/9781644900079-4

In this section, with the help of the mentioned formulas, the fundamental electrochemistry, fuel cell chemistry/electrochemistry, and microbial fuel cell electrochemistry will be discussed.

2. Electrochemistry

2.1 Types of Electrode

An electrode is generally a chemically inert electron conductor in direct contact with an ionic conductor in the form of an electrolyte. Redox-active species in the electrolyte can exchange electrons with the help of the electrode if the electric potential between them provides sufficient driving force. There are three general types of electrodes, the working electrode (WE), the counter electrode (CE), and the reference electrode (RE). Investigations are achieved on the interface of the working electrode by controlling its potential relative to the reference electrode and by measuring the resulting current between the working electrode and the counter electrode, which is described in more detail in section 2.8. The potential at the interface between the reference electrode and the surrounding electrolyte must remain constant to achieve reliable measurements. Potential can be measured only if a current passes and this, in turn, changes the potential. For this reason, the problem is that the measurement changes the system and the three electrode settings are designed to retain this effect at a minimum. A reference and working electrode are connected with a big resistance; thus, almost all current is passed between the counter and working electrode. Moreover, it contains a separate redox system with fast electrode kinetics to quickly adapt to potential changes in the working electrode interface. The ion concentration in the reference electrode must remain constant to provide a constant reference electrode potential. This is accomplished by using a large concentration of potentially detectable ionic species. So, a small flow of current through the reference electrode causes an only negligible turnover. The reference electrode redox system typically comprises a pretty hard soluble salt of metal balanced with a solution consisting cations from this metal and anions of the salt. The potential is detected by the concentration of metal ions in solution. Also, it depends on the concentration of the anions and the solubility product of the hardly soluble salt. Hence, the potential difference between the working and reference electrodes interfaces is at first sensed by the dissolved anions of the reference electrode redox system.

2.2 Potentiostat

The electrical magnitudes between the electrodes can be controlled and measured with a potentiostat system. Moreover, they assist the current flow between working and counter

Enzymatic Fuel Cells
Materials Research Foundations **44** (2019) 73-108

Materials Research Forum LLC
doi: http://dx.doi.org/10.21741/9781644900079-4

electrodes, so that a possible measurement without the flow of current through reference electrode can be provided close to ideal conditions and the external circuit resistance is nearly zero. Fig. 2 shows a simplified potentiostat circuit coupled to an electrochemical device with three electrodes.

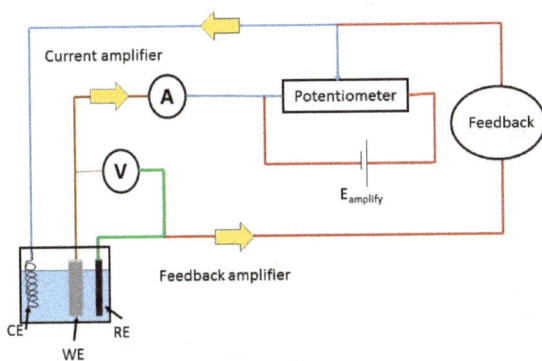

Figure 2. Simplified circuit of a potentiostat, showing the current measurement between WE (red) and CE (green) as well as the potential measurement between WE and RE (blue), while a simplified feedback cycle (orange) enhances current range, inherent accuracy and adjustment speed.

The potentiostat measures the potential difference between working and reference electrodes. It also controls the amplified signal from the feedback circuit. The current is measured between working and counter electrodes and is calculated as the function of the potential difference.

2.3 Electrical Double Layer

Generally, the electric field of an electrode generates a region. In this place, the solution properties differ from the bulk solution because of the re-organization of ions. The concentration differences between the double layer and bulk solution are particularly significant because the response in voltammetric measurements is affected by the double layer thickness. The electrical double layer thickness is the Debye length, κ^{-1}. It varies with the ionic strength of the electrolyte and is usually in the nanometer range. Section 2.9 emphasizes the advantages of reducing the length of the Debye using a high ionic strength electrolyte. The overload on one electrode is counterbalanced by the counter charges in the electrical double layer, and this causes the potential between an ion and the

Enzymatic Fuel Cells
Materials Research Foundations **44** (2019) 73-108

Materials Research Forum LLC
doi: http://dx.doi.org/10.21741/9781644900079-4

electrode to fall with increasing distance on the electrode. As a result, the organization of ions in the electric double layer depends on the distance from the electrode which is why it is divided into several layers with different properties. The number of layers forming the electrical double layer varies between multiple theories modelled according to the respective aspects. Here, the Grahame model is a combination of the Helmholtz and Gouy-Chapman models. It is extended with an additional layer closest to the electrode surface. This layer is called the inner Helmholtz layer [47,48]. Moreover, it accounts for a small number of ionic species that do not have a solvent sheath adsorbed specifically on the electrode surface, as shown in the example of an unsolubilized anion on the positively charged electrode surface in Fig. 3 [21].

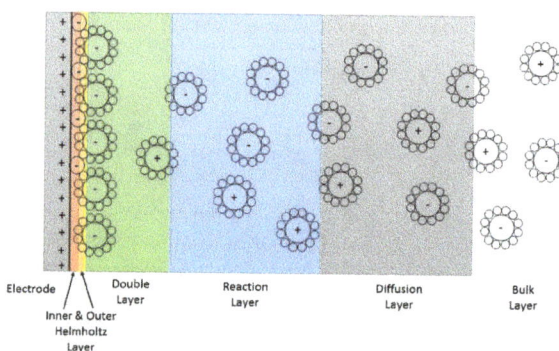

Figure 3. The electric double layer according to the Grahame model is subdivided into Red) inner Helmholtz layer, where not wholly solvated ions are adsorbed explicitly to the electrode surface; yellow) outer Helmholtz layer, which is a monolayer of solvated ions with a distance of their solvation sheath thickness to the electrode; as well as the double (green), reaction (blue) and diffusion (purple) layer, where thermal motion results in lower concentrations of ions.

The specific adsorption to the electrode is not related to the charge of the ion. The thickness of this layer (red) is described by the distance of the plane along the center of the species bound explicitly to the electrode surface. The Helmholtz model allows distinguishing the single layer of solvated ions closest to the electrode (yellow outer Helmholtz layer) from other solvated ions [22]. The thickness of the solvation sheath determines the distance of the nearest solvated ions to the electrode. The width of the outer Helmholtz layer is also described by the plane along the center of these ions. The inner and outer Helmholtz layers form a dense population of ions directly on the

Enzymatic Fuel Cells   Materials Research Forum LLC
Materials Research Foundations **44** (2019) 73-108   doi: http://dx.doi.org/10.21741/9781644900079-4

electrode surface. This causes the potential to decrease linearly from the electrode surface to the outer Helmholtz layer. The Helmholtz layer differs in this aspect from the remaining layers. Because of the potential decreases by the formation of the bulk solution from the Helmholtz layer. This is Chapman, acting against the ordering forces of the electric field. Fig. 4 shows the potential drop with distance from the electrode qualitatively (not to scale).

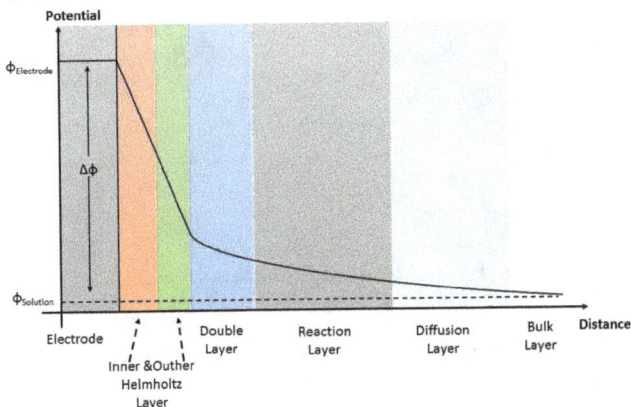

Figure 4. The reduction of the electrode potential ϕ distance by the distance from the electrode is linear in the Helmholtz layer and progresses exponentially outside the Helmholtz layer.

The exponential decline of the potential approaches the potential in the bulk solution [22]. The potential decrease by the distance from the electrode surface is proportional to the net charge accumulation of the ionic species around the electrode. The interface between the electrode and the electrolyte is modeled as a capacitor [20–23–25]. Furthermore, capacitive charging of the electric double layer is frequently called to as capacitive charging of the electrode [20]. The corresponding charge current of the electric double layer is also called the non-Faradaic current. The reason is that electrons are not transferred to the electrode solution interface [25]. For this reason, it should be distinguished from the Faradaic current which is caused by the redox reactions at the electrode in the voltammetric experiments and which represents the background (residual) current. The charging current is proportional to the electrode surface and refers to the rate at which the potential changes [25]. The following relationship shows that the charge current is expressed as a derivative of the capacitor equation [57] concerning time:

Enzymatic Fuel Cells   Materials Research Forum LLC
Materials Research Foundations **44** (2019) 73-108   doi: http://dx.doi.org/10.21741/9781644900079-4

Capacitor equation: $$Q_{dl} = C_{dl}.V \qquad (13)$$

The amount of charge stored in the double layer, Q_{dl}, depends on the double layer capacitance, C_{dl}, and the applied potential V deriving Q_{dl} concerning time gives the charging current ($i_{charging}$):

$$i_{charging} = \frac{\partial Q_{dl}}{\partial t} = C_{dl}.\frac{\partial V}{\partial t} \qquad (14)$$

For this reason, the charge current is proportional to the double layer capacitance, C_{dl}, which is proportional to the electrode surface area. Also, the charge current is proportional to the change in potential over time in the voltammetric experiments, also referred to as scan rate [11].

2.4   Nernst Equation for Electrochemical Equilibrium

Electrochemical experiments usually examine a system by disrupting the electrochemical balance with overpotential and then, measure the response in the form of a current. For the evaluation of these data, the previous equilibrium state must be well known. Forward and reverse reaction arises with the same rate at the chemical equilibrium [26–28] because the chemical potentials of reactants $\mu_{reactants}$ and products $\mu_{products}$ are the same.

$$\sum_{i=1}^{n} \mu_{reactant,i} = \sum_{i=1}^{n} \mu_{product,i} \qquad (15)$$

Where the chemical potential μ_a of any species a is defined as:

$$\mu_a = \mu_a^0 + RT.\ln\frac{[a]}{[\,]^\circ} \qquad (16)$$

With μ_a^o being the temperature and pressure dependent standard chemical potential and $[\,]^\circ$ the standard concentration of 1 mol L^{-1}. The energy term RT relates the molar energy of the given species a to the temperature T of the system through the ideal gas constant, R. Similar to this, the sum of the electrochemical potentials of the oxidized form of a substance $\overline{\mu_{ox}}$ and the electrons used to reduce $\overline{\mu_{e^-}}$ are identical to the electrochemical potential of the reduced form $\overline{\mu_{red}}$ at the electrochemical equilibrium [28]:

$$\sum_{i=1}^{n} \overline{\mu_{ox,i}} + \sum_{i=1}^{n} \overline{\mu_{e^-,i}} = \sum_{i=1}^{n} \overline{\mu_{red,i}} \qquad (17)$$

Enzymatic Fuel Cells   Materials Research Forum LLC
Materials Research Foundations **44** (2019) 73-108   doi: http://dx.doi.org/10.21741/9781644900079-4

The electrochemical potential of the species a is the sum of its chemical potential, μ_a, and its electrical energy, $z_a F \phi$:

$$\overline{\mu}_a = \mu_a + z_a F \phi \tag{18}$$

Where the electrical energy term comes from the charge of species a, z_a, multiplied with both the charge on one mole of electrons, which is the Faraday constant F and the potential of the phase the species a exists in, ϕ. For this reason, the electrochemical potential is defined per mole as same as the chemical potential, as much as it is for the constant temperature and pressure. Electrical energy takes into account the fact that electrons change between two different phases, namely the electrode and the electrolyte [28]. This information is enough to derive an expression to describe the potential difference between an electrode and the surrounding electrolyte for a half-cell of an electrochemical system in an electrochemical balance. Inserting Eq. 17 into Eq. 18 for one redox species that is in electrochemical equilibrium between its oxidized form with the charge Z_{ox} and its reduced form with the charge Z_{red} through the exchange of n electrons per redox process at the electrode gives:

$$(\mu_{ox} + Z_{ox}F\phi_{solution}) + (\mu_{e^-} - nF\phi_{electrode}) = (\mu_{red} + Z_{red}F\phi_{solution}) \tag{19}$$

Where $\phi_{solution}$ is the electric potential of the solution, and $\phi_{electrode}$ is the electric potential of the electrode. Rearrangement of this expression provides a combination of all the chemical potentials on the other side of the equation, and all the contributions made to the electric potential on one side:

$$F[n\phi_{electrode} - \phi_{solution} (Z_{ox} - Z_{red})] = \mu_{ox} + \mu_{e^-} - \mu_{red} \tag{20}$$

The difference between the charge of the oxidized and reduced form of the redox species is equal to the number of electrons exchanged with the electrode during the redox reaction of Z_{ox}-Z_{red}. The potential difference between the electrode and the solution can be expressed as below:

$$\phi_{electrode} - \phi_{solution} = \frac{\mu_{ox} + \mu_{e^-} - \mu_{red}}{nF} \tag{21}$$

Materials Research Forum LLC
doi: http://dx.doi.org/10.21741/9781644900079-4

To relate the above expression to the concentrations of the respective species, Eq. 16 is then added for the chemical potentials:

$$\phi_{electrode} - \phi_{solution} = \frac{\mu^{\circ}ox + \mu e^{-} - \mu^{\circ}red}{nF} + \frac{RT}{nF} \ln \left[\frac{[ox]}{[red]} \right] \tag{22}$$

The chemical potential of electrons does not depend on the electron density. It can, therefore, be merged with the constants μ^{o}_{ox} and μ^{o}_{red} into a new constant $\Delta\mu^{o}$ to give the Nernst equation for the one-half cell:

$$\phi_{electrode} - \phi_{solution} = \frac{\Delta\mu^{\circ}}{nF} + \frac{RT}{nF} \ln \left[\frac{[ox]}{[red]} \right] \tag{23}$$

Although this expression attempts to relate the potential and analyte concentration in theory, it shows in practice that the $\phi_{electrode}$ - $\phi_{solution}$ potential reduction of a single interface between the electrode and solution (one-half cell) cannot be measured. At least one electrode with a special electrode-solution interface potential is required to complete the electrical circuit and to measure the potential difference. This second electrode is the reference electrode (RE) [28]. The working and reference electrodes are separated by the solution with an electrical resistance, $R_{solution}$. This solution resistance contributes to the potential between both electrodes according to Ohm's law (V=i·R) [27–29]. The potential E, measured at the working electrode, is, therefore, the sum of the interfacial potentials ϕ_{int} of both the working and the reference electrodes, plus the potential generation due to the solution resistance:

$$E = \phi_{int,WE} + \phi_{int,RE} + i. R_{solution} \tag{24}$$

reference electrode must have a constant potential to prevent the reference electrode potential decrease from affecting the working electrode readings. The working electrode readings will not be affected by the constant potential of the reference electrode because there is a potential difference between $\phi_{electrode}$ - $\phi_{solution}$, and the solute value is not measured to determine the concentrations. Thus, by utilization of an reference electrode, the Nernst equation for one-half cell (Eq. 23) can be applied to define the measured potential at the working electrode in the equilibrium state [27]:

$$E_{equ} = (\phi_{electrode} - \phi_{solution})_{int,WE} + \phi_{int,RE} + i. R_{solution} = \frac{\Delta\mu^{0}}{nF} + \phi_{int,RE} +$$

$$\frac{RT}{nF} \ln \left(\frac{[ox]}{[red]} \right) + i. R_{solution} \tag{25}$$

The term $\Delta\mu^0/nF + \phi_{int,RE}$ is represented by the standard electrode potential, $E°$ (also written as $E^{\ominus}$). It is detected referenced to the standard hydrogen electrode (SHE). By the use of background electrolyte, the term $i \cdot R_{solution}$ becomes negligible. Finally, this gives the Nernst equation which is practically used to determine the potential for working electrode about standard hydrogen electrode:

$$E_{equ} = E° \frac{RT}{nF} ln\left(\frac{[ox]}{[red]}\right) \tag{26}$$

Electrochemical balance measurements also provide research opportunities such as pH, equilibrium constants, activities, solubility products, free energies, entropies, and enthalpies [27–29].

2.5 Redox Reaction at an Electrode

All energy levels of the electrons in the metallic electrode form an energy continuum known as the conduction band. At absolute zero temperature, the energy maximum of the conduction band is the Fermi level [28–29]. Electrons have more energy at higher temperatures. Under these conditions, the Fermi level is defined as the energy level that is likely to occupy 50 % of an electron. The direction of electron flow at the interface between the electrode and the redox-active species is determined by Fermi level of the electrode according to the lowest molecular orbital (LUMO) [30] and highest occupied molecular orbital (HOMO) [31] energies of redox active species.

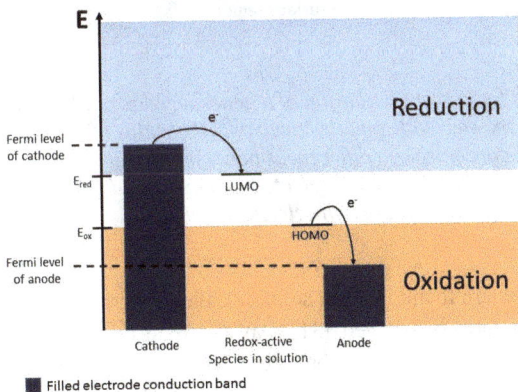

Figure 5. The flow direction of the electrons at the interface between the electrodes and the redox-active species in solution is determined by the Fermi level of the electrode.

Enzymatic Fuel Cells

Materials Research Forum LLC

Materials Research Foundations **44** (2019) 73-108

doi: http://dx.doi.org/10.21741/9781644900079-4

A potentiostat is utilized in electrochemical experiments in most studies. This potentiostat is used to control the Fermi level by adjusting the potential of the working electrode according to the reference electrode. If it should act as a cathode, and if oxidation of the dissolved species on the anode is required, as shown in Fig. 5, the Fermi level of the working electrode should be higher than the LUMO of the species in solution, and vice versa [31].

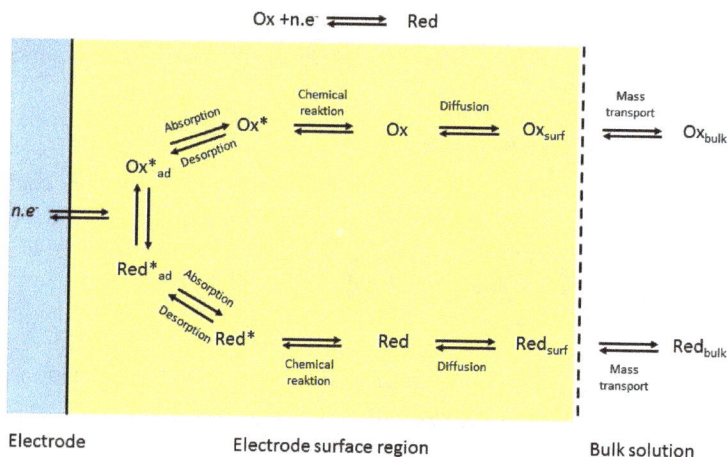

Figure 6. Sources for current limitation of a general redox reaction contain the rate of mass transfer to the electrode and the electrode, the ratios of the necessary chemical reactions, and the concentration differences between the electrode surface and the bulk solution. Redox-active species are detected and measured relative to electron flow through the electrode interface.

Redox-active species are found in a salt solution called electrolyte. The electrode and the electrolyte together form a half cell [31]. An electrode reaches a certain Fermi level relative to the redox potential of the half cell. As a rule, applying an excess negative potential to the electrode induces an electron flow to it. As a result, it also increases the Fermi level. When the Fermi level of the electrolyte is varied, a defined overpotential continuously on the electrode may be applied. If the chemical reaction equation is known,

Enzymatic Fuel Cells   Materials Research Forum LLC
Materials Research Foundations **44** (2019) 73-108   doi: http://dx.doi.org/10.21741/9781644900079-4

the turnover of redox-active species can be calculated because the amount of charge flows and the number of redox reactions is proportional. Monitoring the current reduction during the turnover shows that the redox-active species has been completely converted to a particular oxidation state. Chemical reactions may be required to form redox-active species before its detection at the electrode. It may also limit the current from the electrode and mass transport. Both mass transport and chemical reaction rates affect the concentration differences between the bulk solution and the electrode surface. It also has a significant effect on the electrode kinetics. Current limiting [32] sources including the transfer of n electrons for a general redox reaction at the electrode is summarized in Fig.

If an analyte is assumed to be reduced, this means that the concentration of the oxidized species around the electrode decreases while the concentration of the reduced form increases. If the electron transfer reaction is faster than the diffusion rate, then, these concentration changes become particularly strong in the electric double layer [27–29]. The outcoming concentration gradient of charged species in the electrode surface region conduces to the development of overpotential [33].

In the case that the mass transport into the electrode surface region is not limiting the current, then, the Butler-Volmer equation defines how the measured current, i, differs from the current at zero overpotential, i_0. The difference between bulk concentration, bulk and electrode surface concentration, of redox-active species, is considered as ratios:

$$i = i_0 \cdot \left(\frac{[Red]_0}{[Red]_{bulk}} \exp\left[\frac{(1-\alpha)F\eta}{RT}\right] - \frac{[Ox]_0}{[Ox]_{bulk}} \exp\left[\frac{(1-\alpha)F\eta}{RT}\right] \right) \tag{27}$$

The electron transfer kinetics is denoted by the electron transfer coefficient, α, for the redox reaction between the oxidized species, O_x, and the reduced form, R_{ed} [33]. Mass transport may limit the concentration of a species in the electrode surface region. Thus, it limits the maximum current that can be reached. In such a case, the resulting current is called diffusion limited current. The diffusion limited current is proportional to the mobility and the concentration of the ions [34]. Concerning bio-electricity generation, the diffusion current can be seen as the most optimal performance value [35]. Since achieving this value means that diffusion is limited through the diffusion layer [51–53].

2.6   Consideration of Auxiliary Electrode Processes

The reduction reaction at one electrode requires the oxidation at another electrode and vice versa, to provide the necessary electron flow. For this reason, it is essential to consider the redox reactions on both electrodes to ensure that the electrochemical experiments are correctly established and interpreted. The working electrode can act as either anode or cathode, but in each case, the counter electrode has to fulfil the opposite role to enable the reaction at the working electrode. Therefore, the essential feature of a counter electrode is that it does not limit the current to the working electrode or affect the environment [39–40]. Assuming good mass transport conditions and electron transfer kinetics, the electron flow between the electrodes is limited to the surface of the electrode [41–43]. Since the current flow between working and counter electrodes with different surface areas, one of them will limit the current and cause the potential of this limiting electrode to increase. Therefore, the controlled potential is only the one at the working electrodes. Hence, it should be assured that the current limitation occurs at the working electrode by using a working electrode with a much smaller surface area than the counter electrode has. The rule of thumb for this is that the surface area of the counter electrode should be at least 100 times greater than the surface area of the working electrode. Simple designs from inert and highly conductive noble metals, such as platinum, gold, or glassy carbon further decrease the possibility of interference from the counter electrode. In addition to this, it is essential to confirm that the counter electrode material does not catalyze electrolyte reactions in the applied potential range. In the case that the surface area of the counter electrode is insufficient, the rising potential at the counter electrode would reduce the current. In the worst case, the potential increase at the counter electrode causes the current to flow in the opposite direction concerning the desired direction. According to the setup, such a situation could cause material deposition on the counter electrode instead of the working electrode, or electroactive organisms to lose all benefit from forming a biofilm on an electrode [44].

2.7 Faradaic and Charging Current

Two types of current contributions have been known and these are Faradaic current due to redox processes at the electrode and non-Faradaic current (charging current) due to the double layer capacitance of the electrode. Both types of currents should be distinguished in the assessment of the voltammetric experiments. Charge current flows to a system and the electric field density can be caused by a potential change. It causes the migration of ions in the solution according to the field force. It is also proportional to the surface area of the electrode and the applied sweep rate. Faradaic current depends on the rate of electrons on the surface. The voltage of the electrode comes into play when it provides the necessary redox potential of the species in the electrolyte. The faradic current can be used to examine the concentration of redox-active species, the diffusion coefficient and the redox potential [44]. If these quantities are known, then, the Faradaic current [22] can

Enzymatic Fuel Cells
Materials Research Foundations **44** (2019) 73-108

Materials Research Forum LLC
doi: http://dx.doi.org/10.21741/9781644900079-4

also help to characterize the electrode, as shown in the example of the Randles-Sevcik equation in section 2.10.

2.8 Two, Three and Four Electrode Modes

The three commonly used electrode assemblies, a working electrode, a counter electrode, and a reference electrode, are shown in Fig. 7.

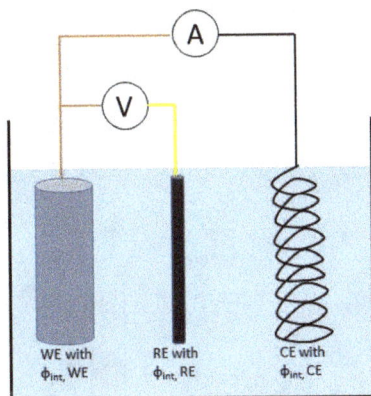

WE with
$\phi_{int, WE}$

RE with
$\phi_{int, RE}$

CE with
$\phi_{int, CE}$

Figure 7. The three electrode settings consisting of the interface potentials ϕ associated with the working electrode (WE), the reference electrode (RE) and the counter electrodes (CE).

The reaction is investigated in the working electrode by controlling the potential against the well-known reference electrode and the fixed electrode. The current flow between working and counter electrodes, thus, allowing the redox reaction at the working electrode to occur by coupling it to the respective reverse reaction. This separation of electrode functions allows analysis of the working electrode containing half cell independently from the changes that might arise in the counter electrode. If necessary, the counter electrode can also fulfil the role of the reference electrode as a pseudo-reference electrode in systems. This approach can only be made in systems with the low current flow and over short measuring times. Conversely, an additional sensing electrode (SE) can be placed to increase the voltage measurement accuracy by completely separating the voltage reading electrodes from the current carrying electrodes [45]. It is presented in more detail in section 2.12.

Enzymatic Fuel Cells   Materials Research Forum LLC
Materials Research Foundations **44** (2019) 73-108   doi: http://dx.doi.org/10.21741/9781644900079-4

2.9   Purpose of Electrolyte

Generally, the electrolyte in a cyclic voltammetry (CV) experiment is mostly redox active species and additional salts as the background electrolyte. There are four reasons to add this background electrolyte. One of them is that the additional salts increase the conductivity of the electrolyte solution. As a result, the resistance of the electrolyte solution is reduced and can be approached negligibly compared to other components of the system. The equation can be simplified for the general potential difference between working and reference electrodes:

$$\Delta V = \varphi_{int,WE} + \varphi_{int,RE} + i.R_{solution} \approx \varphi_{int,WE} + \varphi_{int,RE} \qquad (28)$$

$$i.R_{solution} \approx 0 \qquad (29)$$

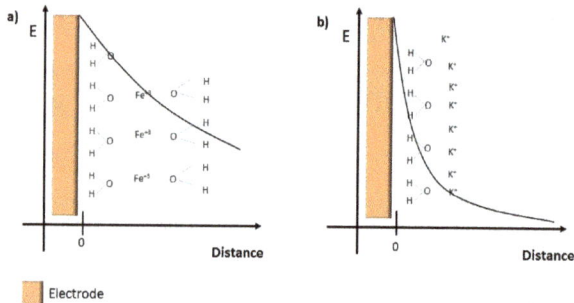

Figure 8. The purpose of the background electrolyte is: a) Without the electrolyte, the electric double layer extends toward the solution, causing the analyte molecules to turn to a static layer around the electrode; b) in the presence of the electrolyte, it falls more rapidly at the potential distance due to charge compensation by the electrolyte.

For this reason, as the conductivity of the electrolyte increases, the test setup will become more complicated because the distance between working and reference electrodes must be precisely defined to determine the contribution of the solution resistor to the potential difference. Furthermore, if the solution comprises of unknown substances, then, specific resistances must be determined separately beforehand. These additional measurement factors bring additional error sources together with them. The high ionic strength of the

Enzymatic Fuel Cells

Materials Research Foundations **44** (2019) 73-108

Materials Research Forum LLC

doi: http://dx.doi.org/10.21741/9781644900079-4

background electrolyte preserves the electric field of the electrode. This electric field causes three problems in the experiments of an aqueous ferricyanide solution, as shown in Fig. 8, both with and without KCI background electrolyte.

According to charge compensation, the molecules in the solution are directed to the electrode surface. Also, the electric stack is stacked to form a static layer around the electrode, called the electrical double layer [41]. The double layer, subdivision and related features are present in more detail in 2.3. When the potential of the electrode is sufficiently large, it causes ions from the solution to be directed to the double layer. The electric double layer thickness κ^{-1}, also known as Debye length, depends on the area of the electric field from the electrode [46]. For this reason, narrowing the double layer is an aim of the electrolyte. The electron exchange between the electrode and redox active species from the solution takes place through tunneling processes. The chance of a tunnelling process decreases exponentially with distance. As a result, for any redox process to take place on the electrode, it is essential that the redox active species is sufficiently close to the electrode surface. Collapsing of the background electrolyte mediated double layer allows the redox active species to enter the electrodes within the tunnel distance. The outer shell electron transfer process can be found in Marcus's Theory [41]. This theory addresses the importance of solvent for Gibbs free energy of activation for each electron transfer [41–43]. The underlying model splits the electron transfer up into the processes of association of electron donor and acceptor, the actual electron exchange between them, and their dissociation. The rate of association and dissociation of donor Do, and acceptor Ac is described with diffusion constants k_{ass} and k_{diss}, and the electron transfer by kinetic rate constants k_{red} and k_{ox}.

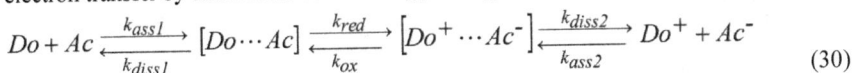

$$Do + Ac \underset{k_{diss1}}{\overset{k_{ass1}}{\rightleftharpoons}} [Do \cdots Ac] \underset{k_{ox}}{\overset{k_{red}}{\rightleftharpoons}} [Do^+ \cdots Ac^-] \underset{k_{ass2}}{\overset{k_{diss2}}{\rightleftharpoons}} Do^+ + Ac^- \qquad (30)$$

If diffusion is controlled more slowly or vice versa kinetically than electron exchange, electron transfer may be diffusion controlled. Calculation of each type of motion (the three types of movements for ions in solution, diffusion, migration, and convection) must be taken into account. The purpose of adding the background electrolyte is to simplify calculations for all different particle movements. Background electrolyte diminishes the impact of migration on general movement. This is obtained by maintaining the electrostatic field of the electrode by increasing the charge density of the solution with the salts of the background electrolyte. As a result, the potential of the electrode is reduced more rapidly with distance. Thus, fewer ions are forced to migrate in a particular direction.

2.10 Cyclic Voltammetry

The cyclic voltammetry is versatile and at the same time one of the widely used electrochemical techniques. In principle, the potentiodynamic electrochemistry, or voltammetry, measures the current evolution while the voltage is changing as described. For material and biofilm studies, one type of voltammetry known as cycling voltammetry is used. It consists of three voltages that define the process of measuring the cyclic voltammogram (CV) with starting potential, first and second vertex potentials. The most positive and negative potential of the cyclic voltammogram is the first and second vertex potential respectively. Therefore, the voltage ranges in which the voltage is increased and decreased (cycled) continuously is set. Ideally, the starting potential does not cause any redox reaction. Thus, the current is zero. However, it can also be set to match the first vertex potential. All three voltage values are determined in a preliminary cyclic voltammogram measurement with the same electrode media as the intended test. The preliminary cyclic voltammogram is carried out over a wide voltage range. The first and second vertex potentials are then fixed to the values that allow all necessary redox reactions to occur but exclude all the others. For this reason, it is essential to optimize the system (electrodes, electrolytes, and dissolved gases in the electrolyte) for each relevant redox reaction. The voltage change (scan) over time for a general first and second vertex potential, E_A and E_B, of a single cyclic voltammogram scan is shown in Fig. 9.

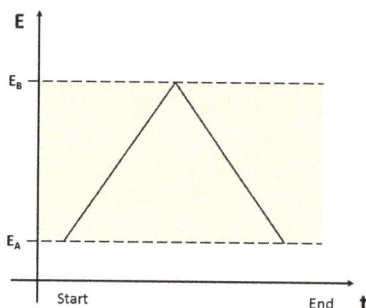

Figure 9. Voltage sweep over time during a CV measurement: The voltage is swept from the first vertex potential EA to the second vertex potential EB and back with a defined sweep/scan rate. The starting potential is usually set in the range from EA to EB at the potential that does not result in current flow.

Materials Research Forum LLC
doi: http://dx.doi.org/10.21741/9781644900079-4

Special equipment allows an excellent linear sweeping. However, sweeping is usually carried out in steps in small increments. Smaller steps provide greater similarity to voltage change to a line. At the same time, however, it increases the measurement time of every single cyclic voltammogram. For this reason, while the smaller voltage steps increase the accuracy of a single cyclic voltammogram, the whole measurement reduces the time resolution. It is, therefore, a general rule that the potential steps must be less than 15th the overall voltage difference to be covered during the cyclic voltammogram. The current resulting from the reduction of the electrolyte appears negative in the cyclic voltammogram, while the oxidation current is positive.

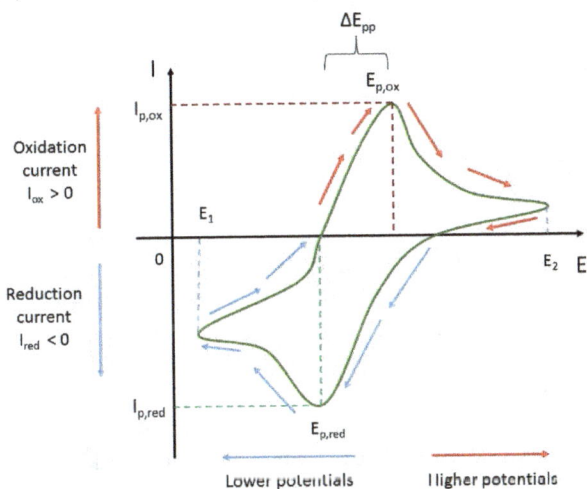

Figure 10. Typical current response of a reversible redox reaction during the potential cycling between the first vertex potential E_1 and second vertex potential E_2 shows an oxidation peak at $E_{p,ox}$ and a reduction peak at $E_{p,red}$ with a peak separation of 57/n mV, where n is the number of electrons exchanged in each redox reaction.

The cyclic voltammogram in Fig. 10 is an example of a single redox reaction with an oxidation potential $E_{p, ox}$ and a reduction potential $E_{p, red}$ at a fixed electrode. Typical properties that define the cyclic voltammogram shape are the peak separation, ΔE_{pp}, the peak size, I_p, and the charge ratio to the Faradaic current. The exchange of these quantities is examined by recording a series of cyclic voltammogram at different scan rates. The peak separation offers information about the reversibility of the redox reaction.

Enzymatic Fuel Cells   Materials Research Forum LLC
Materials Research Foundations 44 (2019) 73-108   doi: http://dx.doi.org/10.21741/9781644900079-4

A constant peak separation at varying scan rates shows a reversible redox reaction with fast electron transfer kinetics. The oxidation and reduction peak of a reversible redox couple are discredited by a potential difference given by:

$$\Delta E_{pp} = \frac{2.218\ RT}{nF} \tag{31}$$

This means that the expected peak separation of a reversible redox process is 57/n mV at 298 K. Contrarily, a non-reversible redox reaction is featured by a scan rate dependent peak separation. The magnitude of the peak current, according to the Randles-Sevcik equation includes information on analyte features, such as C_R concentration and D_R diffusion coefficient:

$$i_{p,solution} = 0.4463nFAC_R \sqrt{\frac{nFD_Rv}{RT}} \tag{32}$$

The peak current amplitude is also detected by the electrode surface area, A, and the scan speed, v. Thus, a redox-active substance of known concentration can be used to identify the active surface area of an electrode; this may differ from the nominal value due to roughness, or it may be due to a broken seal around the conductor [47]. The hysteresis between the current response to the forward scan (E_1 to E_2) and the reverse scan (E_2 to E_1) would also arise in the absence of a redox-active substance and results from the capacitative charging of the electric double layer, as expressed in section 2.3. Applying faster scan rates increases the hysteresis and decreases the peak visibility which is a particular problem in experiments on electrodes with large surface areas. Furthermore, a peak arises in the cyclic voltammogram if the turnover rate of a redox species is faster than its mass transport to the electrode. Consequently, despite the ever-increasing over-potential, the current magnitude reduces after reaching the peak current. The relatively rapid transfer of redox-active species leads to an increase in concentration differences between the bulk solution and the environment of the electrode. The thickness of the diffusion layer, δ_{diff}, can be calculated by the following equations

$$\delta_{diff} \propto \sqrt{\frac{DRT}{vF}} \tag{33}$$

However, a precise value for δ_{diff} depends on the tolerance range of the bulk concentration c_{bulk} because the concentration in the diffusion layer c_0 is asymptotically close to the bulk solution concentration c_{bulk}. Fig. 11 shows the change in c_0 according to

Materials Research Forum LLC
doi: http://dx.doi.org/10.21741/9781644900079-4

the oxidized form (red) of a redox-active oxide as the reduced form (blue) progresses to as the potential erasure proceeds.

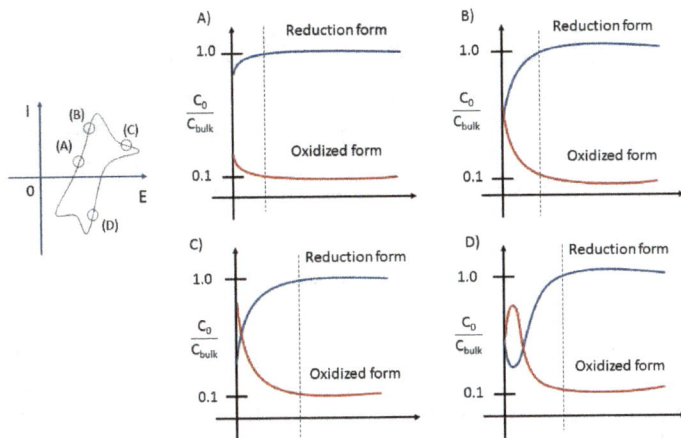

Figure 11. Concentration profiles of the reduced (blue) and the oxidized (red) form of a species as well as the change in diffusion layer thickness D_X at various stages (purple circles) of a CV measurement: A) Small amounts have been oxidized, the diffusion layer is thin and the current is limited by electron transfer kinetics; B) the concentration of reduced form is depleting within the diffusion layer and the current gradient diminishes until a maximum is reached, where the mass transport to the electrode limits the current; C) the diffusion layer extends into the solution to D_{X2} and thereby decreases the concentration gradient between electrode and bulk solution; D) Reduction current replenishes the concentration of reduced form at the electrode and a new double layer will form as soon as the current passes through zero.

Positive current at stage A of the voltammogram demonstrates of oxidation and at the same time consumes the reduced form of the species as the concentration of the oxidized form rises. The current gradient between A and B is detected by electron transfer kinetics as it is sufficient reduction. Growing the overpotential beyond phase B leads to a decrease in the current gradient as the mass transport to the electrode begins to limit the transfer rate. Assuming all other mass transport types are negligible, the peak current is known as mass transport limited current or diffusion limited current. This is usually the case for fixed macro-electrodes in the background electrolyte. The B-phase concentration differences on the electrode begin to reach the bulk solution. The concentration gradient,

Materials Research Forum LLC
doi: http://dx.doi.org/10.21741/9781644900079-4

based on Fick's laws, provides the driving force for the flux of redox species to the electrode in a diffusion-controlled system. For this reason, the lower current at stage C is a result of the lower flux of the species that can be correctly oxidized to the bulk. Step D shows how the reduction reaction during reverse scans replenishes the concentration of the reduced species at the electrode again. In the following forward scan, a new double layer is formed that does not initiate oxidation again and does not repeat the process. This current response is generally observed in voltammetry experiments on fixed electrodes if the diameter is in millimeters and more significant, and if the scan rate is mV s^{-1} and faster. The formation of a peak is useful for identifying and studying specific properties of redox-active species. However, analysis of the solution or electrode kinetics requires a better time resolution and lower charge current to achieve the current limit. This can be achieved by reducing the surface area of the electrode to the minimum, or by carrying the mass more rapidly on the electrode. The smaller surface of the micro-electrodes forms a spherical diffusion profile with enhanced mass transport in the electrode and increases the detection limit due to negligible capacitive charging. Alternatively, the rotating disc electrodes produce a defined laminar flow perpendicular to the electrode surface. Both types of electrodes allow the diffusion coefficients to have better time resolution than with macro-electrodes [47].

2.11 Polarisation and Power Curves

Polarization curves are an essential tool for the analysis and characterization of a (bio)-electrochemical system. They represent the voltage of the cell and the electrodes as a function of the current as obtained by varying the external resistance. A polarization curve can be obtained using variable resistances, but more accurate results are acquired using a potentiostat [47] and can be measured by the applied overpotential, η :

$$H = E - E_{eq} \tag{34}$$

2.12 Four Terminal Sensing

Four-terminal sensing, also known as a 4-point probe or Kelvin sensing, is used to investigate the resistivity of Ti_2AlC without the interference of parasitic resistance from the measurement setup itself. The parasitic resistance is composed of:

1) Resistance from leads and probes, R_{Probe}

2) Construction and surface contamination resistance at the probe to the sample interface, $R_{Contact}$

Enzymatic Fuel Cells   Materials Research Forum LLC
Materials Research Foundations **44** (2019) 73-108   doi: http://dx.doi.org/10.21741/9781644900079-4

3) Spreading resistance due to the current flowing along various paths between the probes, R_{spread} Ti$_2$AlC has previously been reported as an excellent electrical conductor. Four-terminal sensing was employed within this work since parasitic resistance would have caused a significant error in the determination of Ti$_2$AlC resistivity in this case. Fig. 12 indicates the effect of various parasitic resistances on the sample resistance Rsample and two 4-point probe settings used to minimize the measurement interactions.

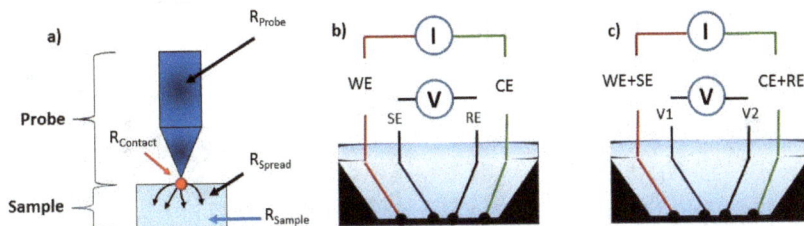

Figure 12. a) Parasitic resistances R_{Probe}, $R_{Contact}$ and R_{Spread} interfering with the determination of the sample resistance R_{Sample}; b) Standard 4-terminal sensing setup with the voltage measuring SE and RE in between the force current carrying WE and CE to minimise lead resistance; c) Specialised 4-terminal sensing setup for the application of an external higher resolution voltmeter with its leads V_1 and V_2.

The decrease of parasitic resistance contribution to the voltage reading is based on the separation of current carrying and voltage sensing electrodes. For this purpose, the voltage is measured with the two inside probes (Fig. 12b). In the case of exceptional low sample resistivity, the voltage magnitude could become so low that an external high sensitivity voltmeter would have to be used. This required the SE to be connected with the working electrode and the reference electrode with the counter electrode to free the voltage sensing probes for the external voltmeter as demonstrated in Fig. 12c [48].

3.   Microbial Fuel Cell (MFC)

Microbial fuel cells use the electrochemical potential generated by the reduction or oxidation of substrates to form an electric current [48–55].

3.1   Types of Microbial Fuel Cell

Generally, microbial fuel cells consist of anodic which can be organic molecules and cathodic compartment which can be oxygen like materials as shown in Figure 13a. There is also proton exchange membrane between these two compartments [50].

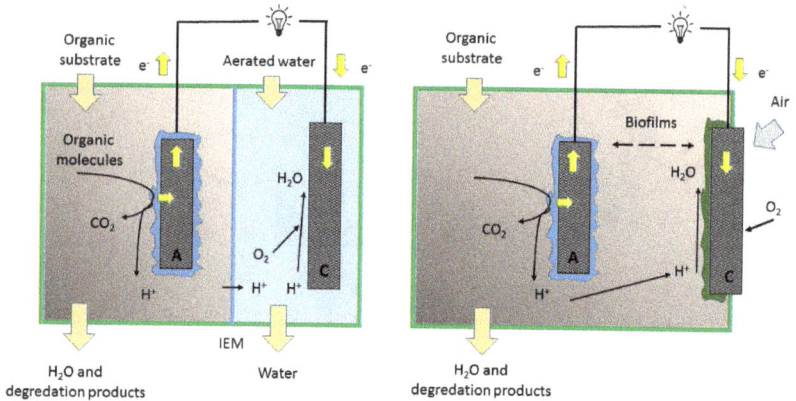

Figure 13. Schematic of a double-chamber (a) and single-chamber (b) MFC.

In microbial fuel cells as shown in Fig. 13, b, the cathode is connected by the reactor to connect the inner system [36].

3.2   Microbial Fuel Cell Principle of Operation

This type of fuel cell is designed to metabolize chemotrophic bacteria to produce electrical energy [50–52–58]. In a microbial fuel cell, bacteria are set in an organic-containing chamber, and a solid conducting anode is substituted for the natural terminal electron acceptors required for their life. Electron transfer in microbial fuel cells can occur by i) direct relationship ii) mediators iii) transferring of electrons.

3.3   Reactions in Microbial Fuel Cell

Gathering of microorganisms adhering to electrode surfaces is called biofilms [57]. The most frequently used substrate for laboratory microbial fuel cells is sodium acetate. Besides, it has been found that microbial fuel cells fed with glucose exhibit anodic biofilms with a greater variety of bacterial populations compared to microbial fuel cells

Enzymatic Fuel Cells
Materials Research Foundations **44** (2019) 73-108

Materials Research Forum LLC
doi: http://dx.doi.org/10.21741/9781644900079-4

fed with acetate. For the glucose-based reactions in a microbial fuel cell can be shown as follows:

Anode Reaction:

$$C_6H_{12}O_6 + 6H_2O \ \rightarrow \ 6CO_2 + 24\,H^+ + 24e^- \qquad E^o = +0.01\ V \qquad (35)$$

Cathode Reaction:

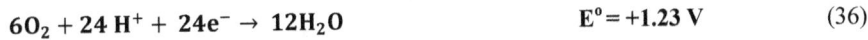

$$6O_2 + 24\,H^+ + 24e^- \rightarrow 12H_2O \qquad E^o = +1.23\ V \qquad (36)$$

The overpotential *(η)* can be defined as the following equation:

$$\zeta = 1 - \eta_a + \eta_c\,/\,\Delta E^o \qquad (37)$$

Reactive oxygen species ensuing from rapid oxygen reduction will create in accelerated ageing and death creating incredibly high oxidative stress on living organisms [103]. Last, but not least, there are also many overcoming problems related to the catalysts which should be very active, reusable, etc. [59–67]. The scientists still demand to get more active catalysts for the fuel cells and also the other applications [68–75], but they are still not able to obtain very efficient catalysts with outstanding performance in terms of stability, durability, activity, reusability, lifetime, etc. [76–84].

4.   Conclusions

In summary, this chapter explains the basis of a fuel cell; fuel cell chemistry; substrates and potentials; electrochemistry; types of electrode; two, three and four electrode modes; purpose of electrolyte; cyclic voltammetry; polarisation and power curves; four-terminal sensing; microbial fuel cell; types of microbial fuel cell; microbial fuel cell principle of operation; reactions in microbial fuel cell; limitations of the electrochemical reaction are described in detail. From these sections, the fuel cells are believed to have an essential impact, especially on the future energy systems. They play a significant role and they will have a potential capability for promising energy systems. Especially, as a clean and renewable energy source, fuel cells have very significant importance.

Materials Research Forum LLC
doi: http://dx.doi.org/10.21741/9781644900079-4

References

[1]   C. Zuo, M. Liu, & M. Liu, Solid Oxide Fuel Cells. Sol-Gel Processing for Conventional and Alternative Energy (Boston, MA: Springer US, 2012), pp. 7–36. https://doi.org/10.1007/978-1-4614-1957-0_2.

[2]   F. M. T. Student, Electricity Generation from Biowaste Based Microbial Fuel Cells. International Journal of Energy, Information and Communications, 1 (2010) 77–92. https://doi.org/10.1016/j.tibtech.2005.04.008.

[3]   Z. Ghassemi & G. Slaughter, Biological Fuel Cells and Membranes. Membranes, 7 (2017) 1–12. https://doi.org/10.3390/membranes7010003.

[4]   A. Baptista, I. Ferreira, & J. Borges, Cellulose-Based Bioelectronic Devices. Cellulose Medical, Pharmaceutical and Electronic Applications (InTech, 2013), pp. 67–82. https://doi.org/10.5772/56721.

[5]   D. E. Edem, C. C. Opara, B. O. Evbuomwan, & B. C. Oforkansi, Effects of Novel Substrates in Electricity Generation in A Mediator-Less Microbial Fuel Cell. Greener Journal of Science, Engineering and Technological Research, 5 (2015) 011–019. https://doi.org/10.15580/GJSETR.2015.1.061715081.

[6]   D. Oladejo, O. O. Shoewu, A. A. Yussouff, & H. Rapheal, Evaluation of Electricity Generation from Animal Based Wastes in A Microbial Fuel Cell. International Journal Of Scientific & Technology Research, 4 (2015) 85–90.

[7]   A. DV & O. TM, Comparative Measure of Electricity Produced from Benthic Mud of FUTA North Gate and FUTA Juction in Akure, Ondo State, Using Microbial Fuel Cell. Innovative Energy & Research, 07 (2018) 1–4. https://doi.org/10.4172/2576-1463.1000180.

[8]   S. D. Purswani, S. S. Atkare, G. Bhumkar, & M. B. Patil, Electricity Generation From Dairy Waste Water through Microbial Fuel Cell Technology. International Journal of Engineering Research and Reviews, 2 (2014) 24–32.

[9]   L. Fan & S. Xue, Overview on Electricigens for Microbial Fuel Cell. The Open Biotechnology Journal, 10 (2016) 398–406. https://doi.org/10.2174/1874070701610010398.

[10]   Z. Samec, Electrochemistry at The Interface Between Two Immiscible Electrolyte Solutions (IUPAC Technical Report). Pure and Applied Chemistry, 76 (2004) 2147–2180. https://doi.org/10.1351/pac200476122147.

[11]   L. Pilon, H. Wang, & A. D'Entremont, Recent Advances in Continuum Modeling of Interfacial and Transport Phenomena in Electric Double Layer Capacitors. Journal of The Electrochemical Society, 162 (2015) A5158–A5178. https://doi.org/10.1149/2.0211505jes.

Materials Research Forum LLC
doi: http://dx.doi.org/10.21741/9781644900079-4

[12]   J.-M. Monier, L. Niard, N. Haddour, B. Allard, & F. Buret, Microbial Fuel Cells: From Biomass (waste) to Electricity. Melecon 2008 - 14th IEEE Mediterr. Electrotech Conference (IEEE, 2008), pp. 663–668. https://doi.org/10.1109/MELCON.2008.4618511.

[13]   A. Parkash, Microbial Fuel Cells: A Source of Bioenergy. Journal of Microbial & Biochemical Technology, 8 (2016) 247–255. https://doi.org/10.4172/1948-5948.1000293.

[14]   Y. Qu, Y. Feng, X. Wang, & B. E. Logan, Use of A Coculture to Enable Current Production by Geobacter Sulfurreducens. Applied and Environmental Microbiology, 78 (2012) 3484–3487. https://doi.org/10.1128/AEM.00073-12.

[15]   M. Rahimnejad, A. Adhami, S. Darvari, A. Zirepour, & S. E. Oh, Microbial Fuel Cell as New Technology for Bioelectricity Generation: A Review. Alexandria Engineering Journal, 54 (2015) 745–756. https://doi.org/10.1016/j.aej.2015.03.031.

[16]   D. Khater, K. M. El-Khatib, M. Hazaa, & R. Y. A. Hassan, Activated Sludge-based Microbial Fuel Cell for Bio-electricity Generation. Journal of Basic and Environmental Sciences, 2 (2015) 63–73.

[17]   R. P. Ramasamy, V. Gadhamshetty, L. J. Nadeau, & G. R. Johnson, Impedance Spectroscopy as A Tool for Non-Intrusive Detection of Extracellular Mediators in Microbial Fuel Cells. Biotechnology and Bioengineering, 104 (2009) 882–891. https://doi.org/10.1002/bit.22469.

[18]   U. Schröder, Anaerobic Biodegradability of Complex Substrates: Performance And Stability at Mesophilic and Thermophilic Conditions. Physical Chemistry Chemical Physics, 9 (2007) 2619–2629. https://doi.org/10.1039/B703627M.

[19]   B. E. Logan, B. Hamelers, R. Rozendal, U. Schröder, J. Keller, S. Freguia, P. Aelterman, W. Verstraete, & K. Rabaey, Microbial fuel cells: Methodology and Technology. Environmental Science and Technology, 40 (2006) 5181–5192. https://doi.org/10.1021/es0605016.

[20]   A. Mallik & B. C. Ray, Evolution of Principle and Practice of Electrodeposited Thin Film: A Review on Effect of Temperature and Sonication. International Journal of Electrochemistry, 2011 (2011) 1–16. https://doi.org/10.4061/2011/568023.

[21]   J. Lin, B. Shi, & Z. Chen, High-Performance Asymmetric Supercapacitors Based on the Surfactant/Ionic Liquid Complex Intercalated Reduced Graphene Oxide Composites. Applied Sciences, 8 (2018) 484. https://doi.org/10.3390/app8040484.

Materials Research Forum LLC
doi: http://dx.doi.org/10.21741/9781644900079-4

[22] J. Varghese, H. Wang, & L. Pilon, Simulating Electric Double Layer Capacitance of Mesoporous Electrodes with Cylindrical Pores. Journal of The Electrochemical Society, 158 (2011) A1106. https://doi.org/10.1149/1.3622342.

[23] P. Cancino, V. Paredes-García, J. Torres, S. Martínez, C. Kremer, & E. Spodine, Electrical Double Layer: Revisit Based on Boundary Conditions. Catalysis Science & Technology., 7 (2017) 4929–4933. https://doi.org/10.1039/C7CY01385J.

[24] K. Bohinc, V. Kralj-Iglič, & A. Iglič, Thickness of Electrical Double Layer. Effect of Ion Size. Electrochimica Acta, 46 (2001) 3033–3040. https://doi.org/10.1016/S0013-4686(01)00525-4.

[25] K. Fushimi, K. I. Takase, K. Azumi, & M. Seo, Current Transients of Passive Iron Observed During Micro-Indentation in Ph 8.4 Borate Buffer Solution. Electrochimica Acta, 51 (2006) 1255–1263. https://doi.org/10.1016/j.electacta.2005.06.016.

[26] G. Sun, A. Thygesen, M. T. Ale, M. Mensah, F. W. Poulsen, & A. S. Meyer, The Significance of The Initiation Process Parameters and Reactor Design for Maximizing The Efficiency of Microbial Fuel Cells. Applied Microbiology and Biotechnology, 98 (2014) 2415–2427. https://doi.org/10.1007/s00253-013-5486-5.

[27] P. M. Biesheuvel, Y. Fu, & M. Z. Bazant, Electrochemistry and Capacitive Charging of Porous Electrodes in Asymmetric Multicomponent Electrolytes. Russian Journal of Electrochemistry, 48 (2012) 580–592. https://doi.org/10.1134/S1023193512060031.

[28] S. Al-Hilli & M. Willander, The Ph Response and Sensing Mechanism of N-Type Zno/Electrolyte Interfaces. Sensors, 9 (2009) 7445–7480. https://doi.org/10.3390/s90907445.

[29] N. Elgrishi, K. J. Rountree, B. D. McCarthy, E. S. Rountree, T. T. Eisenhart, & J. L. Dempsey, A Practical Beginner's Guide to Cyclic Voltammetry. Journal of Chemical Education, 95 (2018) 197–206. https://doi.org/10.1021/acs.jchemed.7b00361.

[30] T.-H. Le, Y. Kim, & H. Yoon, Electrical and Electrochemical Properties of Conducting Polymers. Polymers, 9 (2017) 150. https://doi.org/10.3390/polym9040150.

[31] E. Tran, C. Grave, G. M. Whitesides, & M. A. Rampi, Controlling The Electron Transfer Mechanism in Metal-Molecules-Metal Junctions. Electrochimica Acta, 50 (2005) 4850–4856. https://doi.org/10.1016/j.electacta.2005.04.049.

[32]   J. Babauta, R. Renslow, Z. Lewandowski, & H. Beyenal, Electrochemically Active
       Biofilms: Facts and Fiction. A review. Biofouling, 28 (2012) 789–812.
       https://doi.org/10.1080/08927014.2012.710324.

[33]   S. Minteer & G. Brisard, Physical and Analytical Electrochemistry : The
       Fundamental Core of Electrochemistry. Electrochemical Society Interface, 15
       (2006) 62–65.

[34]   S. A. Ozkan, J.-M. Kauffmann, & P. Zuman, Electroanalysis in Biomedical and
       Pharmaceutical Sciences (Berlin, Heidelberg: Springer Berlin Heidelberg, 2015).
       https://doi.org/10.1007/978-3-662-47138-8.

[35]   D. Pant, G. Van Bogaert, L. Diels, & K. Vanbroekhoven, A Review of The
       Substrates Used in Microbial Fuel Cells (MFCs) for Sustainable Energy
       Production. Bioresource Technology, 101 (2010) 1533–1543.
       https://doi.org/10.1016/j.biortech.2009.10.017.

[36]   R. Kumar, L. Singh, & A. W. Zularisam, Exoelectrogens: Recent Advances in
       Molecular Drivers Involved in Extracellular Electron Transfer and Strategies Used
       to Improve It for Microbial Fuel Cell Applications. Renewable and Sustainable
       Energy Reviews, 56 (2016) 1322–1336. https://doi.org/10.1016/j.rser.2015.12.029.

[37]   V. G. Gude, Wastewater Treatment In Microbial Fuel Cells - An Overview.
       Journal of Cleaner Production, 122 (2016) 287–307.
       https://doi.org/10.1016/j.jclepro.2016.02.022.

[38]   A. V. Sokirko, General Problem of Limiting Diffusion-Migration Currents in A
       System With Ions of Three Arbitrary Charge Numbers. Journal of
       Electroanalytical Chemistry, 364 (1994) 51–62. https://doi.org/10.1016/0022-
       0728(93)02945-E.

[39]   D. Baron, E. LaBelle, D. Coursolle, J. A. Gralnick, & D. R. Bond, Electrochemical
       Measurement of Electron Transfer Kinetics by Shewanella Oneidensis MR-1.
       Journal of Biological Chemistry, 284 (2009) 28865–28873.
       https://doi.org/10.1074/jbc.M109.043455.

[40]   B. M. Setterfield-Price & R. A. W. Dryfe, The Influence of Electrolyte Identity
       upon The Electro-Reduction of CO2. Journal of Electroanalytical Chemistry, 730
       (2014) 48–58. https://doi.org/10.1016/j.jelechem.2014.07.009.

[41]   W. R. Fawcett & M. Opallo, The Kinetics of Heterogeneous Electron Transfer
       Reaction in Polar Solvents. Angewandte Chemie International Edition in English,
       33 (1994) 2131–2143. https://doi.org/10.1002/anie.199421311.

[42]   H. J. Wörner, C. A. Arrell, N. Banerji, A. Cannizzo, M. Chergui, A. K. Das, P.
       Hamm, U. Keller, P. M. Kraus, E. Liberatore, P. Lopez-Tarifa, M. Lucchini, M.

Materials Research Forum LLC
doi: http://dx.doi.org/10.21741/9781644900079-4

Meuwly, C. Milne, J.-E. Moser, U. Rothlisberger, G. Smolentsev, J. Teuscher, J. A. van Bokhoven, & O. Wenger, Charge Migration and Charge Transfer in Molecular Systems. Structural Dynamics, 4 (2017) 061508. https://doi.org/10.1063/1.4996505.

[43]   V. S. Gladkikh, A. I. Burshtein, H. L. Tavernier, & M. D. Fayer, Influence of Diffusion on The Kinetics of Donor-Acceptor Electron Transfer Monitored by The Quenching of Donor Fluorescence. Journal of Physical Chemistry A, 106 (2002) 6982–6990. https://doi.org/10.1021/jp0207228.

[44]   N. K. Bhatti, M. S. Subhani, A. Y. Khan, R. Qureshi, & A. Rahman, Heterogeneous Electron Transfer Rate Constants of Viologen Monocations at A Platinum Disk Electrode. Turkish Journal of Chemistry, 30 (2006) 165–180. https://doi.org/10.1196/annals.1448.047.

[45]   H. T. Nguyen, N. J. Dharan, M. T. Q. Le, N. B. Nguyen, C. T. Nguyen, D. V. Hoang, H. N. Tran, C. T. Bui, D. T. Dang, D. N. Pham, H. T. Nguyen, T. V. Phan, D. T. Dennis, T. M. Uyeki, J. Mott, & Y. T. Nguyen, National Influenza Surveillance in Vietnam, 2006–2007. Vaccine, 28 (2009) 398–402. https://doi.org/10.1016/j.vaccine.2009.09.139.

[46]   K. Asaka & K. Oguro, Bending of Polyelectrolyte Membrane Platinum Composites by Electric Stimuli. Part II. Response Kinetics. Journal of Electroanalytical Chemistry, 480 (2000) 186–198. https://doi.org/10.1016/S0022-0728(99)00458-1.

[47]   M. Q. CHEW, Investigation of Uranium Redox Chemistry and Complexation across The Ph Range by Cyclic Voltammetry, 2013.

[48]   S. Ou, H. Kashima, D. S. Aaron, J. M. Regan, & M. M. Mench, Full Cell Simulation and The Evaluation of The Buffer System on Air-Cathode Microbial Fuel Cell. Journal of Power Sources, 347 (2017) 159–169. https://doi.org/10.1016/j.jpowsour.2017.02.031.

[49]   M. Z. Bazant, Theory of Chemical Kinetics and Charge Transfer based on Nonequilibrium Thermodynamics. Accounts of Chemical Research, 46 (2013) 1144–1160. https://doi.org/10.1021/ar300145c.

[50]   M. Gezginci & Y. Uysal, The Effect of Different Substrate Sources Used in Microbial Fuel Cells on Microbial Community. JSM Environmental Science & Ecology, 4 (2016) 1035.

[51]   A. D. Tharali, N. Sain, & W. J. Osborne, Microbial Fuel Cells in Bioelectricity Production. Frontiers in Life Science, 9 (2016) 252–266. https://doi.org/10.1080/21553769.2016.1230787.

[52]   S. A. Patil, C. Hägerhäll, & L. Gorton, Electron Transfer Mechanisms between Microorganisms and Electrodes in Bioelectrochemical Systems. Bioanalytical Reviews, 1 (2014) 71–129. https://doi.org/10.1007/11663_2013_2.

[53]   P. C. Bogino, M. de las M. Oliva, F. G. Sorroche, & W. Giordano, The Role of Bacterial Biofilms and Surface Components in Plant-Bacterial Associations. International Journal of Molecular Sciences, 14 (2013) 15838–15859. https://doi.org/10.3390/ijms140815838.

[54]   A. E. Franks & K. P. Nevin, Microbial Fuel Cells, A Current Review. Energies, 3 (2010) 899–919. https://doi.org/10.3390/en3050899.

[55]   K. Rabaey & W. Verstraete, Microbial Fuel Cells: Novel Biotechnology for Energy Generation. Trends in Biotechnology, 23 (2005) 291–298. https://doi.org/10.1016/j.tibtech.2005.04.008.

[56]   N. Jayasinghe, A. Franks, K. P. Nevin, & R. Mahadevan, Metabolic Modeling of Spatial Heterogeneity of Biofilms in Microbial Fuel Cells Reveals Substrate Limitations in Electrical Current Generation. Biotechnology Journal, 9 (2014) 1350–1361. https://doi.org/10.1002/biot.201400068.

[57]   L. V. Reddy, S. Pradeep Kumar, & Y.-J. Wee, Microbial Fuel Cells (MFCs)-A Novel Source of Energy For New Millennium. Current Research, Technology and Education Topics in Applied Microbiology and Microbial Biotechnology, (2010) 956–964.

[58]   P. Clauwaert, P. Aelterman, T. H. Pham, L. De Schamphelaire, M. Carballa, K. Rabaey, & W. Verstraete, Minimizing Losses in Bio-Electrochemical Systems: The Road to Applications. Applied Microbiology and Biotechnology, 79 (2008) 901–913. https://doi.org/10.1007/s00253-008-1522-2.

[59]   Y. Yıldız, E. Erken, H. Pamuk, H. Sert, & F. Şen, Monodisperse Pt Nanoparticles Assembled on Reduced Graphene Oxide: Highly Efficient and Reusable Catalyst for Methanol Oxidation and Dehydrocoupling of Dimethylamine-Borane (DMAB). Journal of Nanoscience and Nanotechnology, 16 (2016) 5951–5958. https://doi.org/10.1166/jnn.2016.11710.

[60]   S. Akocak, B. Şen, N. Lolak, A. Şavk, M. Koca, S. Kuzu, & F. Şen, One-Pot Three-Component Synthesis of 2-Amino-4H-Chromene Derivatives by Using Monodisperse Pd Nanomaterials Anchored Graphene Oxide as Highly Efficient and Recyclable Catalyst. Nano-Structures and Nano-Objects, 11 (2017) 25–31. https://doi.org/10.1016/j.nanoso.2017.06.002.

[61]   Z. Daşdelen, Y. Yıldız, S. Eriş, & F. Şen, Enhanced Electrocatalytic Activity and Durability of Pt Nanoparticles Decorated on GO-PVP Hybrid Material for

Methanol Oxidation Reaction. Applied Catalysis B: Environmental, 219 (2017) 511–516. https://doi.org/10.1016/j.apcatb.2017.08.014.

[62] H. Goksu, Y. Yıldız, B. Çelik, M. Yazici, B. Kilbas, & F. Sen, Eco-Friendly Hydrogenation of Aromatic Aldehyde Compounds by Tandem Dehydrogenation of Dimethylamine-Borane In The Presence of A Reduced Graphene Oxide Furnished Platinum Nanocatalyst. Catalysis Science & Technology, 6 (2016) 2318–2324. https://doi.org/10.1039/C5CY01462J.

[63] H. Göksu, Y. Yıldız, B. Çelik, M. Yazıcı, B. Kılbaş, & F. Şen, Highly Efficient and Monodisperse Graphene Oxide Furnished Ru/Pd Nanoparticles for the Dehalogenation of Aryl Halides via Ammonia Borane. ChemistrySelect, 1 (2016) 953–958. https://doi.org/10.1002/slct.201600207.

[64] B. Aday, Y. Yildiz, R. Ulus, S. Eris, F. Sen, & M. Kaya, One-Pot, Efficient And Green Synthesis of Acridinedione Derivatives Using Highly Monodisperse Platinum Nanoparticles Supported with Reduced Graphene Oxide. New Journal of Chemistry, 40 (2016) 748–754. https://doi.org/10.1039/c5nj02098k.

[65] S. Bozkurt, B. Tosun, B. Sen, S. Akocak, A. Savk, M. F. Ebeoğlugil, & F. Sen, A Hydrogen Peroxide Sensor Based on TNM Functionalized Reduced Graphene Oxide Grafted with Highly Monodisperse Pd Nanoparticles. Analytica Chimica Acta, 989 (2017) 88–94. https://doi.org/10.1016/j.aca.2017.07.051.

[66] B. Aday, H. Pamuk, M. Kaya, & F. Sen, Graphene Oxide as Highly Effective and Readily Recyclable Catalyst Using for the One-Pot Synthesis of 1,8-Dioxoacridine Derivatives. Journal of Nanoscience and Nanotechnology, 16 (2016) 6498–6504. https://doi.org/10.1166/jnn.2016.12432.

[67] R. Ayranci, G. Başkaya, M. Güzel, S. Bozkurt, F. Şen, & M. Ak, Carbon Based Nanomaterials for High Performance Optoelectrochemical Systems. ChemistrySelect, 2 (2017) 1548–1555. https://doi.org/10.1002/slct.201601632.

[68] B. Çelik, G. Başkaya, H. Sert, Ö. Karatepe, E. Erken, & F. Şen, Monodisperse Pt(0)/DPA@GO Nanoparticles As Highly Active Catalysts for Alcohol Oxidation and Dehydrogenation of DMAB. International Journal of Hydrogen Energy, 41 (2016) 5661–5669. https://doi.org/10.1016/j.ijhydene.2016.02.061.

[69] E. Erken, I. Esirden, M. Kaya, & F. Sen, A Rapid and Novel Method for The Synthesis of 5-Substituted 1H-Tetrazole Catalyzed by Exceptional Reusable Monodisperse Pt NPs@AC under The Microwave Irradiation. RSC Advances, 5 (2015) 68558–68564. https://doi.org/10.1039/c5ra11426h.

[70] Ö. Karatepe, Y. Yıldız, H. Pamuk, S. Eris, Z. Dasdelen, & F. Sen, Enhanced Electrocatalytic Activity and Durability of Highly Monodisperse Pt@PPy–PANI

Materials Research Forum LLC
doi: http://dx.doi.org/10.21741/9781644900079-4

Nanocomposites as A Novel Catalyst for The Electro-Oxidation of Methanol. RSC Advances, 6 (2016) 50851–50857. https://doi.org/10.1039/C6RA06210E.

[71] E. Erken, H. Pamuk, Ö. Karatepe, G. Başkaya, H. Sert, O. M. Kalfa, & F. Şen, New Pt(0) Nanoparticles as Highly Active and Reusable Catalysts in the C1–C3 Alcohol Oxidation and the Room Temperature Dehydrocoupling of Dimethylamine-Borane (DMAB). Journal of Cluster Science, 27 (2016) 9–23. https://doi.org/10.1007/s10876-015-0892-8.

[72] F. Sen, Y. Karatas, M. Gulcan, & M. Zahmakiran, Amylamine Stabilized Platinum(0) Nanoparticles: Active and Reusable NanocatalystIn The Room Temperature Dehydrogenation of Dimethylamine-Borane. RSC Advances, 4 (2014) 1526–1531. https://doi.org/10.1039/c3ra43701a.

[73] S. Eris, Z. Daşdelen, Y. Yıldız, & F. Sen, Nanostructured Polyaniline-rGO Decorated Platinum Catalyst with Enhanced Activity and Durability for Methanol Oxidation. International Journal of Hydrogen Energy, 43 (2018) 1337–1343. https://doi.org/10.1016/j.ijhydene.2017.11.051.

[74] Y. Yıldız, S. Kuzu, B. Sen, A. Savk, S. Akocak, & F. Şen, Different Ligand Based Monodispersed Pt Nanoparticles Decorated with rGO as Highly Active and Reusable Catalysts for The Methanol Oxidation. International Journal of Hydrogen Energy, 42 (2017) 13061–13069. https://doi.org/10.1016/j.ijhydene.2017.03.230.

[75] Y. Yıldız, H. Pamuk, Ö. Karatepe, Z. Dasdelen, & F. Sen, Carbon Black Hybrid Material Furnished Monodisperse Platinum Nanoparticles as Highly Efficient and Reusable Electrocatalysts for Formic Acid Electro-Oxidation. RSC Advances, 6 (2016) 32858–32862. https://doi.org/10.1039/C6RA00232C.

[76] E. Erken, Y. Yıldız, B. Kilbaş, & F. Şen, Synthesis and Characterization of Nearly Monodisperse Pt Nanoparticles for C_1 to C_3 Alcohol Oxidation and Dehydrogenation of Dimethylamine-borane (DMAB). Journal of Nanoscience and Nanotechnology, 16 (2016) 5944–5950. https://doi.org/10.1166/jnn.2016.11683.

[77] B. Çelik, E. Erken, S. Eriş, Y. Yildiz, B. Şahin, H. Pamuk, & F. Sen, Highly Monodisperse Pt(0)@AC NPs as Highly Efficient And Reusable Catalysts: The Effect of The Surfactant on Their Catalytic Activities in Room Temperature Dehydrocoupling of DMAB. Catalysis Science and Technology, 6 (2016) 1685–1692. https://doi.org/10.1039/c5cy01371b.

[78] B. Çelik, S. Kuzu, E. Erken, H. Sert, Y. Koşkun, & F. Şen, Nearly Monodisperse Carbon Nanotube Furnished Nanocatalysts as Highly Efficient and Reusable Catalyst for Dehydrocoupling of DMAB and C1 to C3 Alcohol Oxidation.

Materials Research Forum LLC
doi: http://dx.doi.org/10.21741/9781644900079-4

International Journal of Hydrogen Energy, 41 (2016) 3093–3101.
https://doi.org/10.1016/j.ijhydene.2015.12.138.

[79] G. Baskaya, İ. Esirden, E. Erken, F. Sen, & M. Kaya, Synthesis of 5-Substituted-
      1H-Tetrazole Derivatives Using Monodisperse Carbon Black Decorated Pt
      Nanoparticles as Heterogeneous Nanocatalysts. Journal of Nanoscience and
      Nanotechnology, 17 (2017) 1992–1999. https://doi.org/10.1166/jnn.2017.12867.

[80] B. Sen, S. Kuzu, E. Demir, S. Akocak, & F. Sen, Polymer-Graphene Hybride
      Decorated Pt Nanoparticles as Highly Efficient and Reusable Catalyst for The
      Dehydrogenation of Dimethylamine–Borane at Room Temperature. International
      Journal of Hydrogen Energy, 42 (2017) 23284–23291.
      https://doi.org/10.1016/j.ijhydene.2017.05.112.

[81] E. Demir, B. Sen, & F. Sen, Highly Efficient Pt Nanoparticles and f-MWCNT
      Nanocomposites Based Counter Electrodes for Dye-Sensitized Solar Cells. Nano-
      Structures & Nano-Objects, 11 (2017) 39–45.
      https://doi.org/10.1016/j.nanoso.2017.06.003.

[82] S. Eris, Z. Daşdelen, & F. Sen, Investigation of Electrocatalytic Activity and
      Stability of Pt@f-VC Catalyst Prepared by In-Situ Synthesis for Methanol
      Electrooxidation. International Journal of Hydrogen Energy, 43 (2018) 385–390.
      https://doi.org/10.1016/j.ijhydene.2017.11.063.

[83] S. Eris, Z. Daşdelen, & F. Sen, Enhanced Electrocatalytic Activity and Stability of
      Monodisperse Pt Nanocomposites for Direct Methanol Fuel Cells. Journal of
      Colloid and Interface Science, 513 (2018) 767–773.
      https://doi.org/10.1016/j.jcis.2017.11.085.

[84] B. Sen, A. Şavk, & F. Sen, Highly Efficient Monodisperse Pt Nanoparticles
      Confined in the Carbon Black Hybrid Material for Hydrogen Liberation. Journal
      of Colloid and Interface Science, 520 (2018) 112–118.
      https://doi.org/10.1016/j.jcis.2018.03.004.

Enzymatic Fuel Cells
Materials Research Foundations **44** (2019) 109-130

Materials Research Forum LLC
doi: http://dx.doi.org/10.21741/9781644900079-5

Chapter 5

Biological Fuel Cell Applications

Binti Srivastava[1], Madhu Khatri[1], Shailendra Kumar Arya[1]*

[1]Department of Biotechnology, University Institute of Engineering and Technology, Panjab University, Chandigarh, India

skarya_kr@yahoo.co.in

Abstract

Biological fuel cells are the bio-electro chemical cells that convert chemical energy into electrical energy utilising microbes as an active biocatalyst. These are termed as green electricity generation devices because of their sustainable and eco-friendly nature, energy storage capability, and are also cost-effective. This chapter provides detailed information on biological fuel cell applications in the field of antibiotics removal from wastewater, power generation, and removal of soil contaminants such as arsenic and iron oxide, biosensors for monitoring toxic compounds, controlling denitrification, treating wastewater arising from food industry, municipal waste, sanitary wastes, sewage and simultaneously generating electricity.

Keywords

Biological Fuel Cell, Electricity Generation, Waste Water Treatment, Biosensors, Eco-Friendly, Biocatalyst, Microbes

Contents

Materials Research Forum LLC
doi: http://dx.doi.org/10.21741/9781644900079-5

1.   Introduction

Since the past half century, there is an increased demand for energy and water supplies due to an enormous rise in the human population. The primary source of energy on earth is the sun and human beings have introduced millions of ways in order to utilize energy from it [1, 2]. The emission of toxic gases due to the continuous use of fossil fuels and the depletion of natural reserves had led researchers to develop a renewable form of energy i.e. wind and sun. Regrettably, because of their sporadic nature, the renewable forms of technologies are not able to maintain the balance of the electricity supply. The wastewater treatment plants generate waste biomass that causes problems to the society as well as the environment. Therefore, researchers are focused on developing green technologies in order to meet the demand of people, in respect of production, the capability of energy storage, sustainability, electricity, environment friendliness and economically viability [1]. One recent green or renewable form of technology found by researchers are microbial fuel cells (MFCs) that utilize solid waste and then convert it into a renewable energy reserve. Microbial fuel cells also known as biological fuel cells are considered as the most promising technology that utilizes waste water and converts it into electricity [3].

Microbial fuel cells have the ability to use microbial populations as a catalyst and produce electricity from an extensive variety of living and non-living substances. MFCs can be of various types, such as the single chamber, double chamber, mediatorless, and mediated MFCs. In the single chamber MFC, there is no boundary between the anode and the cathode, due to which the diffusion of dissolved oxygen has an antagonistic effect on the anodic electrode which inhibits the direct electron transfer from the biofilm to the surface of the anode. Inoculation of the cell along with substrate, vitamins and minerals, in a fed-batch single, chambered MFC, microbial fuel cell battery is improved and the

Enzymatic Fuel Cells   Materials Research Forum LLC
Materials Research Foundations **44** (2019) 109-130   doi: http://dx.doi.org/10.21741/9781644900079-5

cell voltage is steadily increased to the maximum voltage value and after series of repetitions, it finally plunged to its lowest value [4]. There are certain mass transport limitations in microbial fuel cells, for which these are scaled up close to the pilot plants. Flat plate reactors are considered as the most sophisticated design to avoid the mass transport limitation. For better MFC performance, the interior of anodes is structured with low electrode spacing and high retention times to improve mass transport. Carbon fabric electrodes when compared to carbon felts, paper or foils, showed better results for mass transfer [5].

Since MFC need no metal catalyst at the anode surface therefore as compared to other fuel cells, it is distinctive. As an alternative, electrogenic bacteria cover the anode surface and act as the biocatalyst for the fuel cell and oxidize the organic substrate. Electricity can be produced in MFC when electrogenic type bacteria in absence of oxygen, reduce iron by accepting electrons. Proteins (large molecules), acetate and glucose (simple molecule) and complex fusions of organic matter that are found in sludge, food waste and domestic wastewater are a biodegradable organic matter that has the ability to be used as a substrate to generate electricity in microbial fuel cell [6]. Recent advances in microbial fuel cell technologies include the pathways for electron transfer, the alignment for the anodic chamber and the formation of biofilm and the intra-biofilm transport processes [7].

Microbial fuel cells due to their tremendous characteristics can be utilized for various purposes, such as, in the brewery and domestic wastewater treatment, electricity generation, bio-remediation, biosensors and purification of food waste, etc. [4]. This chapter provides detailed information about all the applications related to biological fuel cells.

2.   Types of fuel cells

2.1   Fuel cells

Fuel cells are chemical devices that have the ability to directly convert inherent chemical energy present in fuels to electrical energy. The different types of fuel cells are schematically represented in Figure 1. Fuel cells do not get affected by the thermodynamic limitations as defined by the Carnot efficiency because the most conservative power generation in terms of intermediary steps of heat production is avoided [8]. Fuel cells promise to provide high-efficiency power generation and less environmental pollution. In a fuel cell setup, anode compartment is continuously being supplied by the fuel while cathode compartment is fed with the oxidant, for example,

Materials Research Forum LLC

doi: http://dx.doi.org/10.21741/9781644900079-5

oxygen. The electricity is generated due to the electrochemical reactions that occur at the anode (where oxidation occurs) and cathode (where reduction occurs) via the electrolytes [8] as represented in Figure 2.

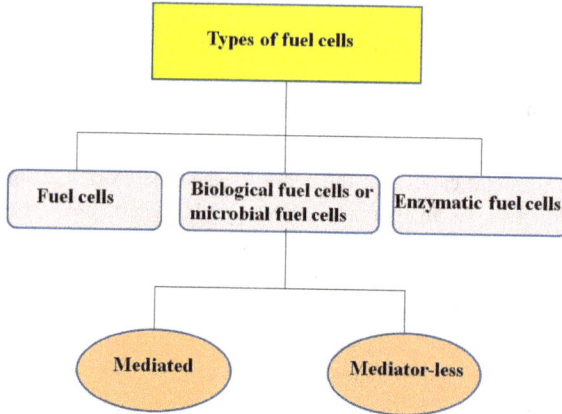

Figure 1: A flowchart representation of different fuel cells.

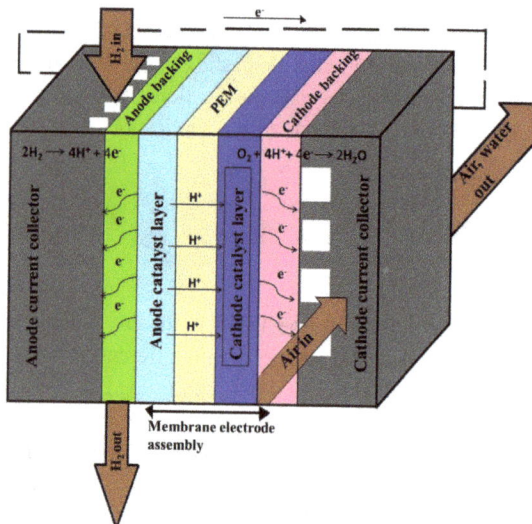

Figure 2: A schematic representation of principle and operation of the chemical fuel cell.

Enzymatic Fuel Cells   Materials Research Forum LLC
Materials Research Foundations **44** (2019) 109-130   doi: http://dx.doi.org/10.21741/9781644900079-5

2.2 Biological fuel cells or Microbial fuel cells

Biological fuel cells, also known as bio-electrochemical systems [9] work on the principle of oxidation-reduction reactions where with the help of microbial metabolism energy from organic wastes is generated [10]. MFCs consist of an anaerobic anode chamber and aerobic cathode chamber. Both the chambers are separated by proton exchange membrane (PEM) or salt bridge which is made up of a nafion or ultrex membrane. Microbes in the anode chamber are used as a biocatalyst for the production of bio-electricity [11]. Microbial fuel cells can be classified into two types, such as (1) Mediated microbial fuel cells that require certain kind of mediators, for example, humic acid, electron shuttle, methyl blue for transferring electrons to the electrodes. Most of the mediators used in MFCs are costly and toxic in nature [12], (2) Mediator less microbial fuel cells that do not require any kind of mediators, rather these utilize bacteria to transfer electrons to the electrode [13]. The phenomenon occurs in two steps. Firstly, in the anode chamber, micro-organisms oxidize organic or inorganic wastes and generate electrons and protons. Electrons produced by the bacteria are transferred to the anode electrode and then flow to the cathode through an external circuit whereas, protons are transferred through the PEM membrane [14]. Secondly, in the aerobic cathode chamber, protons and electrons combine to form water and electricity through the reduction of oxygen [4] as shown in following reactions (1) and (2) and represented in Figure 3.

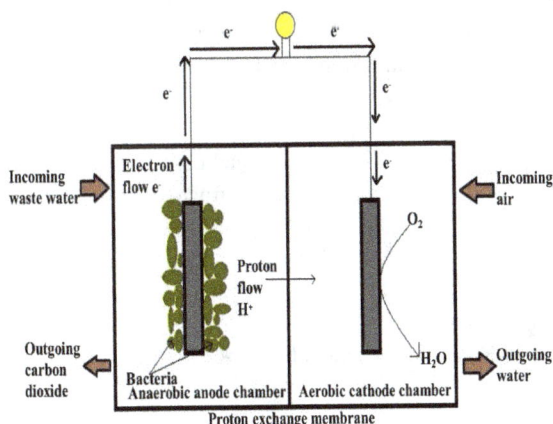

Anode: $CH_3COO^- + 2H_2O \rightarrow 2CO_2 + 7H^+ + 8e^-$ (1)

Cathode: $O_2 + 4H^+ + 4e^- \rightarrow 2H_2O$ (2)

Figure 3: A schematic diagram showing the working principle of the biological fuel cell.

Materials Research Forum LLC
doi: http://dx.doi.org/10.21741/9781644900079-5

Substrates used in MFCs provide energy for the bacterial cell. Some of the most common substrates used in MFCs include acetate, lactate, glucose, sucrose, xylose, sodium fumarate, phenol and starch, etc. [15]. Substrates influence total power density and columbic efficiency. Proton exchange membranes are favoured for use in MFCs because of their high permeability and low internal resistance [10]. Some micro-organisms such as *Geobacter spp., Aeromonas hydrophila, Pseudomonas aeruginosa, Shwenalla spp.* and *Rhodobacter sphaeroide* have been most frequently tested for power generation in MFCs [15]. The greatest advantage of biological fuel cells over anaerobic digestion is that it generates power from wastes without producing gases like hydrogen and methane and therefore considered as the cleanest form of technology [16].

2.3   Enzymatic Fuel Cells

Enzymatic fuel cells are the subclasses of fuel cells that deal with wastewater treatment and electricity generation [17]. In terms of power densities, a volumetric size and wide range of applications (from small scale to large scale), enzymatic and microbial fuel cells differ. Enzymatic fuel cells consist of highly refined wild-type engineered enzymes that are used for driving microelectronic systems, costly environmental remediation and hygiene systems while microbial fuel cells consist of bacteria and yeast that are of low cost [18, 19, 20]. Yet, microbial fuel cells are considered a most eco-friendly method for generating electricity from waste and purification of wastewater. In contrast, enzymatic fuel cells are used as portable power sources for wearable and implantable electronic devices due to their superior power density and compactness. These fuel cells seem to transmit greater potential as battery-free power elucidations [21]. Recent reviews are available for further information on important topics such as the progression of bioinspired strategies for the improvement of high surface area bio-electrodes and the electrical wiring of the enzymes. Remarkable inventions on high surface area electrodes include the porous magnesium oxide templated carbon and carbon nanotube yarn textile electrodes [22, 23]. The purpose of enzymatic fuel cells is mainly to generate electricity by the oxidation of saccharides such as glucose, fructose, pyruvate, hydrogen, methanol and ethanol at the anode and reduction of oxygen to water at the cathode. Now a day, enzymatic fuel cells had gathered special attention to medical devices and also transportable electronic devices for example sensors, mobile phones, GPS systems and laptops [21]. The most recent development of biofuels deals with the hydrogen/oxygen enzymatic fuel cells [24] as represented in Figure 4.

Materials Research Forum LLC
doi: http://dx.doi.org/10.21741/9781644900079-5

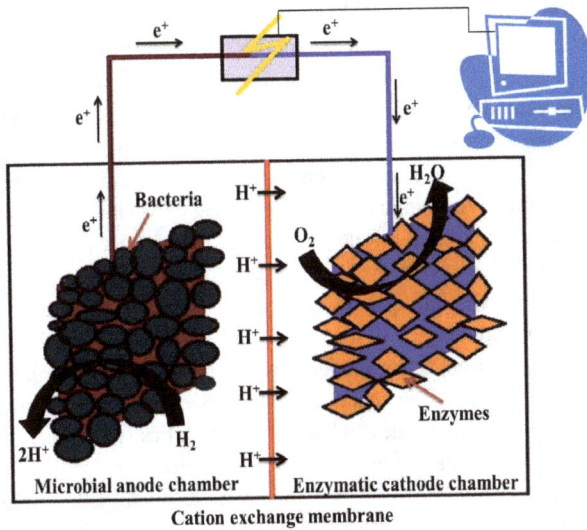

Figure 4: Enzymatic fuel cell with microbial anode and enzymatic cathode chambers.

3. Biological Fuel Cell Applications

The pictorial applications of biological fuel cells as shown in Figure 5 have been discussed in detail in the following paragraphs of this chapter.

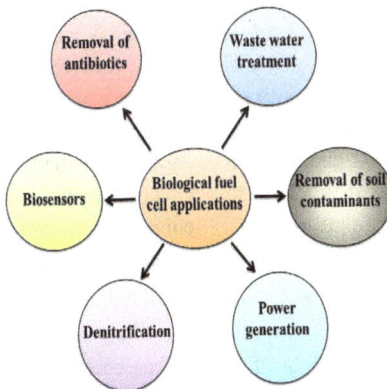

Figure 5: Pictorial representation of major applications of biological fuel cells.

3.1   Antibiotics removal from wastewater

In order to promote the growth of animals, antibiotics are broadly consumed worldwide as herbs and are considered as one of the most vital chains of medications in human as well as in veterinary treatment [25, 26]. Due to population growth and global economic, antibiotic ingestion worldwide is considerably increased, which is about 90,000-190,000 tons per year [27]. The frequent consumption of antibiotics leads to possible side effects on human beings. Antibiotic tetracycline causes yellowing teeth, gastrointestinal problem and liver damage, even at low concentrations and produces allergic reaction if a person is extremely sensitive. Moreover, intake of antibiotics for long periods leads to drug resistance [26]. Multiple aquatic environments have been frequently detected with the antibiotics that are applied in the prevention and treatment of diseases [25]. So, aquatic environment, as well as its related environmental issues and public health problems, remain a serious environmental problem. Therefore, it is crucial to adopt suitable measures to resolve issues related to antibiotics pollution, especially in animal wastewater. Traditional, as well as other biological methods like ultrasonic cavitation, chlorination, adsorption, advanced oxidation technology, membrane process and the electromagnetic treatment, are being used to remove antibiotics from wastewater but these are not ideal for practical applications [26]. Unpredictably, microbial fuel cell, an eco-friendly technology, has the capability to convert organic matter by utilizing the microorganism as a biocatalyst into bioelectricity [27]. It has been revealed that MFCs can excellently degrade sulphamethoxazole and hence refractory pollutants in the environment could also be degraded by microbial fuel cell [28]. Through the advancement of microbial fuel cell machinery, the investigations about MFCs have gained much publicity concerning the structures, microorganisms and electrodes [29]. But then again, there is partial evidence existing about eliminating pollutants like antibiotics by the microbial fuel cell, which involved diversities of microorganism and produce electricity [26]. Antibiotics like sulphonamide and aureomycin are cleared into the surroundings because most of them are difficult to be engrossed and putrefied by animals [30]. Roxithromycin and norfloxacin are one of the kinds of antibiotics where roxithromycin have a long shelf-life, also recalcitrant in the environment, while yearly manufactures of norfloxacin are used for breeding pigs. With the liberation of sewage, the antibiotics mentioned above-containing wastewaters can pollute the environment [31].

To determine the effect of extra pollution created by antibiotics, biological fuel cells/microbial fuel cells comprising of biofilm has been successfully introduced along with the wastewater and replicated with animal wastewater as the substrate once the anode created and occupied with the bacterial biofilm [26]. The relationship between the

Enzymatic Fuel Cells   Materials Research Forum LLC
Materials Research Foundations **44** (2019) 109-130   doi: http://dx.doi.org/10.21741/9781644900079-5

performance of biological fuel cell and antibiotics were established by various approaches under optimal conditions. These involved: (1) the deletion situation of antibiotic during the process of the biological fuel cell; (2) the possible deletion technology of antibiotic by biological fuel cell devices; and (3) the removal efficacy of contamination by biological fuel cell devices [26]. Figure 6 illustrates antibiotics removal from wastewater by the anaerobic self-electrolysis process. By using supernatant of mixtures of sludge and inoculation done with animal wastewater in biological fuel cell/microbial fuel cell, electricity was produced successfully. The results showed that without the addition of antibiotics, the internal resistance, stable voltage, and the maximum power density of anaerobic self-electrolysis i.e. ASE-116 and ASE-112 were 29.38 U, 0.565 V and 5.82 Wm^{-3} and 28.06 U, 0.574 V, and 5.78 Wm^{-3}, respectively [26]. Furthermore, with the addition of sulfadimidine, norfloxacin, aureomycin and roxithromycin into the reactors, microbial fuel cell performance was reduced from 0.51 V to 0.41 V, whereas due to the reduced concentration of antibiotics the output voltage was augmented. Even though the deletion of pollutants such as total nitrogen, chemical oxygen demand and nitrate nitrogen exhibited an irrelevant performance, the deletion of total phosphorus and ammonia nitrogen was significantly improved. At the same time, the deletion efficacy of norfloxacin, aureomycin, roxithromycin and sulfadimidine showed 100% through LC-MS analysis. These results showed that antibiotics displayed significantly inhibitions for electricity performance but improved the quality of water simultaneously [26].

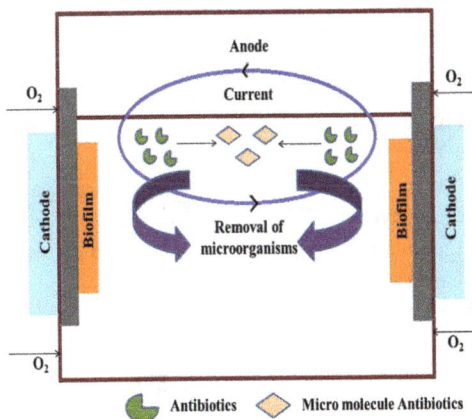

Figure 6: The schematic illustration of the removal of antibiotics via anaerobic self-electrolysis.

Enzymatic Fuel Cells   Materials Research Forum LLC
Materials Research Foundations **44** (2019) 109-130   doi: http://dx.doi.org/10.21741/9781644900079-5

3.2 Power generation

As described earlier, biological fuel cells are the microbial fuel cell that harnesses the natural metabolism of microorganism as the biocatalyst at the anode chamber for the production of bio-electricity by the conversion of chemical energy into electrical energy [32]. Fuels are oxidized in the anode compartment by the catalyst consisting of redox enzymes, whereas reduction of oxygen occurs in the cathode compartment [33]. Conventional fuel cell technology for environmental devices such as monitors, sensors and conventional battery power showed less availability for replacement, maintenance and narrow lifespan as compared to biological fuel cells that produce energy from wastewater and thus biological fuel cells possess a great advantage over conventional fuel cell technology [33, 34]. Earlier, the microbial fuel cell was tested in the marine environment and the experiment systems were arranged in the benthic zone. In such type of experiment setup, the anode was submerged in sediment and indigenous bacteria create a population that respires using the anode as an electron acceptor, thus producing current [33]. The sediments being organic-rich provide sustenance and a miscellaneous bacterial population adapted to its environment; utilization of naturally occurring microbes that eventually populate the anode buried inside the sediment to form a biofilm that characteristically creates higher power densities than just seeded anodes covered with cultured isolates [33].

In most of the microbial fuel cells, cathode reaction is a limiting factor in power output. Platinum a costly noble metal was used as a catalyst in conventional fuel cell technology on which effective oxygen reduction occurred [35]. This limitation overcome by biological catalysts by the integration of multi-copper oxidase (oxygen reduction catalyst) to the cathode of an MFC tends to increase the theoretical electrical potential of the system as compared to metal catalyst [33]. A speculative electromotive force of 1.1 volts was achieved when a microbial anode was combined with a laccase cathode which classically required the use of mediators as a vehicle to transport the electrons from the cathode, in result causing over potential losses [36]. On the other hand, it is advantageous to use a biological fuel cell that comprises of a microbial anode and enzymatic cathode that employs direct electron transfer requires no mediators and possesses simple design with great operating potential [37]. Biological fuel cells to be operated in a marine environment with non-indigenous biological practices need to overcome certain design constraints that limit power output. Other limiting factor includes the structural design and materials of the electrode. At present, the cathode part of the biological fuel cell, comprised of multi-copper oxidases, such as bilirubin oxidase and laccase, that catalyse oxidation-reduction reaction might decrease the power output in saline water since they

Enzymatic Fuel Cells   Materials Research Forum LLC
Materials Research Foundations **44** (2019) 109-130   doi: http://dx.doi.org/10.21741/9781644900079-5

are inhibited by chloride. The problem of chloride inhibition can overcome by the active direct electron transfer between the surface of electrodes and multi-copper oxidase [38].

For the testing potential of power output in the marine environment, a hybrid biological fuel cell having microbial anode for oxidation of lactate and enzymatic cathode for reduction of oxygen was constructed. The anode was "pre-seeded" with a microbial model organism named *Shewanella oneidensis* DSP-10 which was cultured in research laboratory medium and then via the help of silica sol-gel, was fixed on a carbon felt electrode in order to catalyse fuel cell reactions [33]. The enzyme bilirubin oxidase acted as an electrocatalyst, covered the cathode, and with the help of heterobifunctional crosslinker was fixed to an electrode comprised of carbon nanotube, and then further stabilised with a coating of silica sol-gel. The hybrid biological fuel cell was maintained in raw seawater for more than a week which encouraged the performance of the system as shown in Figure 7.

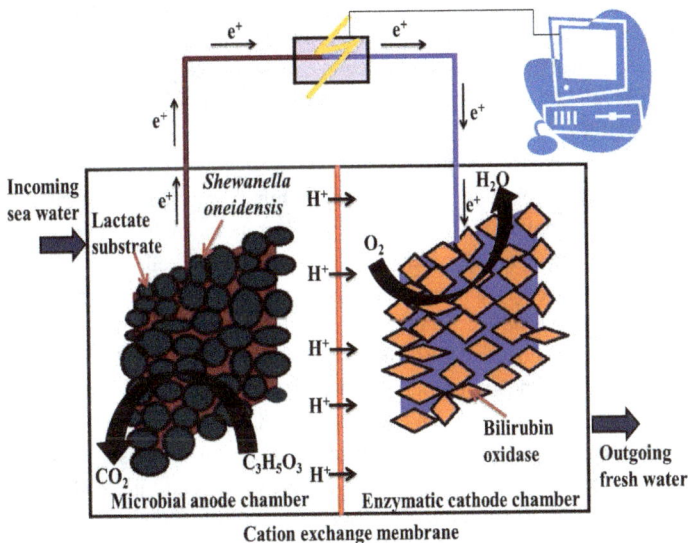

Figure 7: Schematic representation of hybrid biological fuel cell with microbial anode chamber and enzymatic cathode chamber used for power generation.

Upon the release of the enzymatic cathode, power output increased due to the carbon nanotubes. The mean open circuit voltage of the anodic cell in seawater was in the range

Materials Research Forum LLC
doi: http://dx.doi.org/10.21741/9781644900079-5

of "-0.43 to -0.45 Volts (vs. Ag/AgCl)" while that of the cathodic cell was "0.477±0.005 Volts". The overall mean operating potential provided by the hybrid biological fuel cell was "0.786±0.040 Volts" which indicated that the hybrid biological fuel cell sustained a consistent open circuit voltage in the laboratory which was higher than 0.7 Volt for 9 days and electrocatalytic activity was sustained open surroundings tests for more than 24 hours [33].

3.3 Microbial fuel cell acts as a biosensor

Biosensors are biological devices that aim to determine toxic pollutants. Carbon monoxide is one of the most dangerous gases. Various human chemical activities and by-products of power plants are considered extremely harmful for the human health as well as deteriorate the environment [39, 40]. Therefore, it is necessary to detect the leakage of gases from time to time in order to protect human beings and workplaces which are affected by carbon monoxide. During the past years, a lot of efforts have been directed toward the development of electrochemical, optical and semiconductors carbon monoxide sensors. These sensors being highly sensitive can detect the presence of carbon monoxide even in very small concentrations ranging from 1 to 100 mg/L [41, 42]. But, to determine high concentrations of carbon monoxide emerging from steel mill factories or from incomplete combustion is still a challenge. Moreover, noble metal fabrications which are too expensive are required by the maximum number of conventional sensors. Also, numerous conventional sensors operate at temperatures higher than 100 °C and thus demand excess energy [43]. Recently, many microbial fuel cells have gained importance as attractive biosensors to determine high concentrations of carbon monoxide. Various applications can be monitored by microbial fuel cell-based biosensors, such as chemical or biological oxygen demand [26], reduction of heavy metals (Cu^{2+}) and nitric oxide, volatile fatty acids and microbial activity [43]. Toxic pollutants like Cu^{2+} and formaldehyde can be detected by microbial fuel cells based biosensors [44]. Even though carbon monoxide is consumed by *Clostridium* species as a substrate, still most of the microorganisms are affected by it. As a result, the activity of bio-anode in the microbial fuel cell may possibly get inhibited by the presence of carbon monoxide concentration that can be indicated by voltage drop [43].

A novel microbial fuel cell-based biosensor with anodic biofilm (also termed as anode exoelectrogens) has been developed to determine toxic carbon monoxide at high concentration and to interpret the relationship between carbon monoxide concentration and voltage as illustrated in Figure 8.

Enzymatic Fuel Cells
Materials Research Foundations **44** (2019) 109-130

Materials Research Forum LLC
doi: http://dx.doi.org/10.21741/9781644900079-5

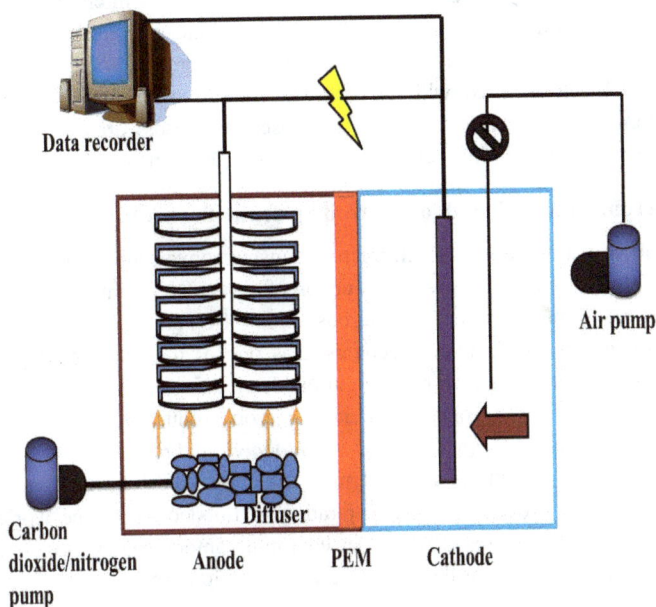

Figure 8: Schematic representation of bioelectrical carbon monoxide sensor.

In order to eliminate the reduction of electricity production and inhibition of bacterial activity at the anodic surface by carbon monoxide, an advanced electrochemically dynamic biofilm on the anode was exposed to carbon monoxide gas at different concentrations [43]. Carbon monoxide concentration was determined by the fall in voltage that showed the toxicity of carbon monoxide which was 0.8 to 24 mV in the range of 10 and 70% of carbon monoxide concentration and no further voltage was increased when the carbon monoxide concentration was increased to above 70%. A comparative linear relationship (R^2=0.987) was established between voltage and carbon monoxide concentration and the recommended response time of biosensor was 50 minutes to 1 hour [43].

Because of the simple design and operation, biosensors have shown favourable insights. It can be applied to the developed remote place where there is an absence of electricity because these sensors can operate without the addition of electricity. The bioelectrical sensor developed here shows favourable insights. For example, ignition amenities arise due to high carbon monoxide production can be detected by microbial fuel cell-based biosensors. Besides this, biosensors possess great advantages as these can operate under room temperature, economically feasible, do not require any electricity if single chamber microbial fuel cell is hired with air cathode and do not require expensive metals distinguishing components. Although MFC based biosensors are favourable, still the working operation, selectivity, stability and sensitivity are some challenges [43].

3.4 Microbial fuel cell used for controlling denitrification

Denitrification is a process in which microorganisms convert inorganic nitrogen present in soil into elemental nitrogen gas i.e. reduction of nitrate (NO_3^-) into gaseous compounds like nitrite (NO_2^-), nitric oxide (NO), nitrous oxide (N_2O) and finally nitrogen (N_2) [45]. Atmospheric nitrous oxide (N_2O) recognised as one of the greenhouse gases causing global warming found as a major source in paddy rice fields [46]. Denitrification is the main reason behind the nitrogen deficiency in crop production. Denitrifying bacteria, such as *Pseudomonas, Alcaligens, Bacillus, Thiobacillus, Vibrio, Agrobacterium,* etc. in absence of oxygen, use nitrate as the terminal electron acceptor which gets reduced to nitrogen during the process of their respiration in flooded soils. The worldwide food industry is dependent on fossil fuels. An enormous amount of energy is required for fixing nitrogen fertilizer from the unlimited atmospheric nitrogen. It has been estimated that 3500 m^3 of natural gas manufactures one metric ton of anhydrous ammonia that contains 82% of nitrogen. Also, 645 m^3 of natural gas is consumed for 150 kg per hectare nitrogen fertilizer in the form of ammonia. Since nitrogen fertilizer is essential for supporting the worldwide population, it is crucial to discover efficient ways in order to use nitrogen fertilizer to overcome forthcoming food scarcities [45]. Various conventional methods for example irrigation management, production of leguminous crops and crop rotation have been planned and applied to improve nitrogen holding efficiency [47, 48]. But these methods are ineffective during a change in climatic conditions. A universally accepted technology, the microbial fuel cell has been employed for saving nitrogen in agricultural fields since denitrification has caused a huge amount of nitrogen loss.

As stated earlier, microbial fuel cells are bio-electrochemical devices that convert chemical energy into electrical energy. Redox reaction is the main principle behind the power generation in a microbial fuel cell. This principle of the microbial fuel cell can be applied to flooded rice soil in order to control denitrification rates to overcome the losses of applied nitrogen [45]. In MFC, there is a transfer of electrons after decomposition of organic matter from the anode to the cathode where there is a reduction of oxygen (terminal electron acceptor) to water in order to generate electricity coupled with redox modifications. Electricity could be generated in flooded soils. The soil near the water surface and the soil several centimetres below the flooded soil surface are aerobic and anaerobic respectively [45]. When connected through shielded wires, the electrical potential could be created among these two types of soils [49]. During microbial metabolism, the electrons are released through the oxidation of organic matter which can be utilized for generating electricity as shown in Figure 9.

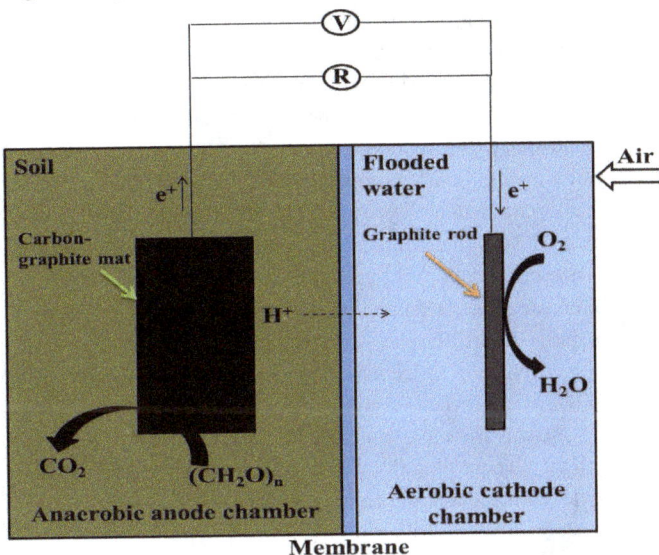

Figure 9: Schematic explanation microbial fuel cell theory in flooded soil.

Due to this, the accessibility of electrons gets reduced for half reactions of nitrate so that denitrification is suppressed. In the process of denitrification of flooded rice fields, inorganic nitrogen species containing oxygen such as nitrate and nitrite gain electrons generated due to the oxidation of organic matter via microorganisms. Classic redox reactions performed by microbes in the absence of oxygen related to nitrate respiration are shown in the equations. (3) and (4).

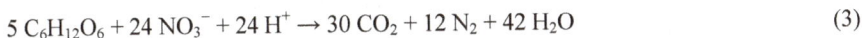

$$5\,C_6H_{12}O_6 + 24\,NO_3^- + 24\,H^+ \rightarrow 30\,CO_2 + 12\,N_2 + 42\,H_2O \tag{3}$$

$$C_{12}H_{22}O_{11} + 9.6NO_3^- + 9.6H^+ \rightarrow 12CO_2 + 4.8N_2 + 15.8H_2O \tag{4}$$

To determine nitrogen losses in soils, planting pots with gas chamber experiments were set up under three conditions, i.e. microbial fuel systems, microbial fuel systems with an externally applied voltage and non-microbial fuel system as a control [45]. The same amount of nitrogen fertilizer was applied to the triplicate system and flooded with automatic irrigation. Inorganic nitrogen concentration, soil redox potential and nitrous oxide flux, in soil pore water was intermittently examined. The results showed that the soil redox potentials of both microbial fuel systems with and without externally applied voltage were considerably higher than that of non- microbial fuel systems, whereas nitrous oxide flux levels were considerably lower than that of non-microbial fuel systems [45]. The carbon/nitrogen ratio, as well as the redox potential in a microbial fuel cell with externally applied voltage and microbial fuel cells without externally applied voltage, showed consistency in results. Rice growth is vital during the reproductive stage period because of the enhanced availability of nitrogen fertilizer during this period [45]. The application of externally applied voltage on limiting the soil redox potential and nitrous oxide flux was not elucidated. To overcome the effect of externally applied voltage through the soil, considerable steps required taken to improve the design of anode. With constant nitrous oxide flux differences, the results showed that microbial fuel systems have higher nitrogen retention efficiency in pore water. The non-microbial fuel cell was estimated with 6.6% of denitrified nitrogen while that of microbial fuel cell systems was estimated at 2.3%. It was therefore concluded that microbial fuel cells applications can suppress denitrification, control soil redox potential and nitrogen losses in addition to lower nitrous oxide (a greenhouse gas) emissions [45].

3.5   Wastewater treatment

For crop production, industrial wastewater and sewage can be alternative sources in the area with limited freshwater [50]. In agricultural fields, fresh water supplied represents more than 70% of the total water use in the world [51]. Even though wastewater helps to overcome water scarcity and wastewater disposal, still it results in the reduction of crop yield and quality and also possesses hazard to the environment. Degradation of organic matter by aerobic microbes for consumption of oxygen limits oxygen availability in soil due to overloading of organic matter [51]. Soil health and blocking of soil pores can be harmed by the overloading of organic carbon due to the limitation of oxygen availability [52]. The growth of biofilms results in the reduction of porosity of soils [53]. Also, inorganic nutrients (phosphorus and nitrogen) and heavy metals damage the soil as well as crops. Hence, before using wastewater in agricultural lands, it should be treated. A number of technologies (conventional wastewater treatment plants, lagoon ponds, membrane filtration, constructed wetlands, membrane bioreactors and etc.) can be used to treat wastewater before applying to agricultural lands. But the disadvantage of using these methods includes high investment cost and the requirement of large land areas [54]. Therefore, the wastewater should be treated by microbial fuel cells since they are economically feasible and overcome energy shortages.

A microbial fuel cell is renewable green technology that treats wastewater and at the same time generates electricity while leaving low concentrations of nutrients in the wastewater. A microbial fuel cell works on the principle of a redox reaction. It consists of two electrodes, namely, anodic and cathodic electrodes respectively. Anode chamber is continuously supplied with wastewater such as wastes from the brewery and food industry, sewage waste, sanitary wastewater and municipal wastes demonstrated in [51]. Anode consists of bacteria that decompose organic matter present in wastewater and produce electrons and protons. An electron through the external circuit reaches the cathode while protons transfer internally through cation exchange membrane that is only selective for protons. The electrons and protons recombine together at cathode chamber, where oxygen (electron acceptor) is reduced to water, generating electricity.

Conclusion

This chapter demonstrates various biological fuel cell applications by utilizing microorganisms as an active biocatalyst to treat wastewater arising from various sources and simultaneously electricity production. MFCs play a significant role in the removal of antibiotics from wastewater for improving the quality of water. MFCs eliminate toxic soil pollutants such as heavy metals and inorganic solvents to improve soil fertility and improve crop yield. For better performance, measures must be taken to overcome the low power density of microbial fuel cell and to improve the quality of the electrodes by the introduction of new engineered bio-electrodes.

References

[1]   P. Ratajczak, M.E. Suss, F. Kaasik, F. Beguin, Carbon electrodes for capacitive technologies, Energy Storage Mater. 16 (2019) 126–145. https://doi.org/10.1016/j.ensm.2018.04.031

[2]   C.M. Fernandez-Marchante, Y. Asensio, L.F. Leon, J. Villasenor, P. Canizares, J. Lobato, M.A. Rodrigo, Thermally-treated algal suspensions as fuel for microbial fuel cells, J. Electroanal. Chem. 814 (2018) 77–82. https://doi.org/10.1016/j.jelechem.2018.02.038

[3]   N. Zhao, Y. Jiang, M. Alvarado-Morales, L. Treu, I. Angelidaki, Y. Zhang, Electricity generation and microbial communities in microbial fuel cell powered

by macroalgal biomass, Bioelectrochemistry 123 (2018) 145–149. https://doi.org/10.1016/j.bioelechem.2018.05.002

[4]   D.Z. Khater, K.M. El-Khatib, R.Y.A. Hassan, Effect of vitamins and cell constructions on the activity of microbial fuel cell battery, J. Genet. Eng. Biotechnol. (2018) 0-4.

[5]   T. Krieg, J.A. Wood, K.M. Mangold, D. Holtmann, Mass transport limitations in microbial fuel cells: Impact of flow configurations, Biochem. Eng. J. 138 (2018) 172-178. https://doi.org/10.1016/j.bej.2018.07.017

[6]   A.R. Ruslan, V.M. Vadivelu, Nitrite pre-treatment of dewatered sludge for microbial fuel cell application, J. Environ. Sci. (2018) 2–9.

[7]   J. Hu, Q. Zhang, D. Lee, H.H. Ngo, Feasible use of microbial fuel cells for pollution treatment, Renew. Energy. 129 (2018) 824-829. https://doi.org/10.1016/j.renene.2017.02.001

[8]   K. Scott, An introduction to microbial fuel cells, Microbial Electrochemical and Fuel Cells, Elsevier Ltd., UK, 2016, pp. 3-27. https://doi.org/10.1016/B978-1-78242-375-1.00001-0

[9]   A. Ebrahimi, G.D. Najafpour, D. Youse, Performance of microbial desalination cell for salt removal and energy generation using different catholyte solutions, Desalination 432 (2018) 1–9. https://doi.org/10.1016/j.desal.2018.01.002

[10] M. Li, M. Zhou, X. Tian, C. Tan, C.T. McDaniel, D.J. Hassett, T. Gu, Microbial fuel cell (MFC) power performance improvement through enhanced microbial electrogenicity, Biotechnol. Adv. 36 (2018) 1316–1327. https://doi.org/10.1016/j.biotechadv.2018.04.010

[11] D.R. Lovley, Bug juice : harvesting electricity with microorganisms, Nature Rev. Microbiol. 4 (2006) 497–508. https://doi.org/10.1038/nrmicro1442

[12] G. Delaney, H.P. Bennetto, J.R. Mason, S.D. Roller, J.L. Stirling, C.F. Thurston, Electron-transfer coupling in microbial fuel cells. 2. Performance of fuel cells containing selected microorganism-mediator-substrate combinations, J. Chem. Tech. Biotechnol. 34 (2008) 13–27. https://doi.org/10.1002/jctb.280340104

[13] B. Kim, Dynamic effects of learning capabilities and profit structures on the innovation competition, Optim. Control Appl. Meth. 20 (1999) 127–144. https://doi.org/10.1002/(SICI)1099-1514(199905/06)20:3<127::AID-OCA650>3.0.CO;2-I

[14] H.M.M. Selim, A.M. Kamal, D.M.M. Ali, R.Y.A. Hassan, Bioelectrochemical systems for measuring microbial cellular functions, Electroanal. 29 (2017) 1498-1505.

[15] I. Ulusoy, A. Dimoglo, Electricity generation in microbial fuel cell systems with *Thiobacillus ferrooxidans* as the cathode microorganism, Int. J. Hydrogen Energy. 43 (2018) 1171–1178. https://doi.org/10.1016/j.ijhydene.2017.10.155

[16] T. Krieg, F. Enzmann, D. Sell, J. Schrader, D. Holtmann, Simulation of the current generation of a microbial fuel cell in a laboratory wastewater treatment plant, Appl. Energy. 195 (2017) 942–949. https://doi.org/10.1016/j.apenergy.2017.03.101

[17] C. Santoro, C. Arbizzani, B. Erable, I. Ieropoulos, Microbial fuel cells: From fundamentals to applications: a review, J. Power Sources. 356 (2017) 225–244. https://doi.org/10.1016/j.jpowsour.2017.03.109

[18] G. Pasternak, J.Greenman, I. Ieropoulos, Self-powered, autonomous biological oxygen demand biosensor for online water quality monitoring, Sensors and Actuators B. Chem. 244 (2017) 815–822. https://doi.org/10.1016/j.snb.2017.01.019

[19] T. Tommasi, G. Lombardelli, Energy sustainability of microbial fuel cell (MFC): A case study, J. Power Sources. 356 (2017) 438-447. https://doi.org/10.1016/j.jpowsour.2017.03.122

[20] I.A. Ieropoulos, A. Stinchcombe, I. Gajda, S. Forbes, I. Merino-jimenez, G. Pasternak, D. Sanchez, J. Greenman, Pee power urinal – microbial fuel cell technology field trials in the context of sanitation, Environ. Sci. Water Res. Technol. 2 (2016) 336-343. https://doi.org/10.1039/C5EW00270B

[21] S. Cosnier, A.J. Gross, F. Giroud, M. Holzinger, Beyond the hype surrounding biofuel cells: What's the future of enzymatic fuel cells? Curr. Opin. Electrochem. (2018) 1–8.

[22] C.H. Kwon, S. Lee, Y. Choi, J.A. Lee, S.H. Kim, H. Kim, D. Lima, M.E. Kozlov, R.H. Baughman, G.M. Spinks, G.G. Wallace, S.J. Kim, High-power biofuel cell textiles from woven biscrolled carbon nanotube yarns, Nature Comm. 5 (2014) 1–7. https://doi.org/10.1038/ncomms4928

[23] S. Tsujimura, K. Murata, W. Akatsuka, Exceptionally high glucose current on a hierarchically structured porous carbon electrode with "wired" flavin adenine dinucleotide - dependent glucose dehydrogenase, J. Am. Chem. Soc. 136 (2014) 14432-14437. https://doi.org/10.1021/ja5053736

[24]   L. Xu, F.A. Armstrong, Pushing the limits for enzyme-based membrane-less hydrogen fuel cells – achieving useful power and stability, RSC Adv. 5 (2014) 3649–3656. https://doi.org/10.1039/C4RA13565B

[25]   N. Kemper, Veterinary antibiotics in the aquatic and terrestrial environment, Ecolo. Indi. 8 (2008) 1–13. https://doi.org/10.1016/j.ecolind.2007.06.002

[26]   Y. Zhou, N. Zhu, W. Guo, Y. Wang, X. Huang, P. Wu, Z. Dang, X. Zhang, J. Xian, Simultaneous electricity production and antibiotics removal by microbial fuel cells, J. Environ. Manage. 217 (2018) 565–572. https://doi.org/10.1016/j.jenvman.2018.04.013

[27]   R. Wise, Leading articles Antimicrobial resistance : priorities for action, J. Antimicro. Chemotherapy 49 (2002) 585–586. https://doi.org/10.1093/jac/49.4.585

[28]   L. Wang, Y. Liu, J. Ma, F. Zhao, Rapid degradation of sulphamethoxazole and the further transformation of 3-amino-5-methylisoxazole in a microbial fuel cell, Water Res. 88 (2016) 322–328. https://doi.org/10.1016/j.watres.2015.10.030

[29]   J. Wei, P. Liang, X. Huang, Recent progress in electrodes for microbial fuel cells, Bioresour. Technol. 102 (2011) 9335–9344. https://doi.org/10.1016/j.biortech.2011.07.019

[30]   A. Spielmeyer, B. Breier, K. Groißmeier, G. Hamscher, Elimination patterns of worldwide used sulfonamides and tetracyclines during anaerobic fermentation, Bioresour. Technol. 193 (2015) 307–314. https://doi.org/10.1016/j.biortech.2015.06.081

[31]   H. Kim, Y. Hong, J. Park, V.K. Sharma, S. Cho, Sulfonamides and tetracyclines in livestock wastewater, Chemosphere 91 (2013) 888–894. https://doi.org/10.1016/j.chemosphere.2013.02.027

[32]   L. Hao, B. Zhang, M. Cheng, C. Feng, Effects of various organic carbon sources on simultaneous V (V) reduction and bioelectricity generation in single chamber microbial fuel cells, Bioresour. Technol. 201 (2016) 105–110. https://doi.org/10.1016/j.biortech.2015.11.060

[33]   G. Strack, H.R. Luckarift, S.R. Sizemore, R.K. Nichols, K.E. Farrington, P.K. Wu, P. Atanassov, J.C. Biffinger, G.R. Johnson, Power generation from a hybrid biological fuel cell in seawater, Bioresour. Technol. 128 (2013) 222–228. https://doi.org/10.1016/j.biortech.2012.10.104

[34]   Y. Gong, S.E. Radachowsky, M. Wolf, M.E. Nielsen, P.R. Girguis, C.E. Reimers, Benthic microbial fuel cell as direct power source for an acoustic modem and

seawater oxygen / temperature sensor system, Environ. Sci. Technol. 45 (2011) 5047–5053. https://doi.org/10.1021/es104383q

[35] C.H. Kjaergaard, J. Rossmeisl, J.K. Norskov, Enzymatic versus inorganic oxygen reduction catalysts : comparison of the energy levels in a free-energy scheme, Inorg. Chem. 49 (2010) 3567–3572. https://doi.org/10.1021/ic900798q

[36] O. Schaetzle, F. Barriere, U. Schroder, An improved microbial fuel cell with laccase as the oxygen reduction catalyst, Energy Environ. Sci. 2 (2009) 96–99. https://doi.org/10.1039/B815331K

[37] S.R. Higgins, C. Lau, P. Atanassov, S.D. Minteer, M.J. Cooney, Hybrid biofuel cell : microbial fuel cell with an enzymatic air-breathing cathode, ACS Catal. 1 (2011) 994–997. https://doi.org/10.1021/cs2003142

[38] C. Vaz-dominguez, S. Campuzano, R. Olaf, M. Pita, M. Gorbacheva, S. Shleev, V.M. Fernandez, A.L. De Lacey, Laccase electrode for direct electrocatalytic reduction of O_2 to H_2O with high-operational stability and resistance to chloride inhibition, Biosensors and Bioelectronics 24 (2008) 531–537. https://doi.org/10.1016/j.bios.2008.05.002

[39] M. Diender, A.J.M. Stams, D.Z. Sousa, Pathways and bioenergetics of anaerobic carbon monoxide fermentation, Frontiers Microbiol. 6 (2015) 1–18. https://doi.org/10.3389/fmicb.2015.01275

[40] A. Paliwal, A. Sharma, M. Tomar, V. Gupta, Carbon monoxide (CO) optical gas sensor based on ZnO thin films, Sensors Actuators B. Chem. 250 (2017) 679-685. https://doi.org/10.1016/j.snb.2017.05.064

[41] M. Shojaee, S. Nasresfahani, M.H. Sheikhi, Hydrothermally synthesized Pd loaded SnO_2/partially reduced graphene oxide nanocomposite for effective detection of carbon monoxide at room temperature, Sensors Actuators B. Chem. 254 (2017) 457-467. https://doi.org/10.1016/j.snb.2017.07.083

[42] S. Kundu, R. Sudarshan, M. Narjinary, Pd impregnated gallia: Tin oxide nanocomposite - An excellent high temperature carbon monoxide sensor, Sensors Actuators B. Chem. 254 (2018) 437-447. https://doi.org/10.1016/j.snb.2017.07.094

[43] S. Zhou, S. Huang, Y. Li, N. Zhao, H. Li, I. Angelidaki, Y. Zhang, Microbial fuel cell-based biosensor for toxic carbon monoxide monitoring, Talanta 186 (2018) 368–371. https://doi.org/10.1016/j.talanta.2018.04.084

[44] J. Sun, G.P. Kingori, R. Si, D. Zhai, Z. Liao, D. Sun, T. Zheng, Y. Yong, Microbial fuel cell-based biosensors for environmental monitoring : a review, Water Sci. Technol. 71 (2015) 801–809. https://doi.org/10.2166/wst.2015.035

[45]   T. Ranatunga, K. Hiramatsu, T. Onishi, Controlling the process of denitrification
       in flooded rice soils by using microbial fuel cell applications, Agric. Water Manag.
       206 (2018) 11–19. https://doi.org/10.1016/j.agwat.2018.04.041

[46]   H. Akiyama, K. Yagi, X. Yan, Direct N_2O emissions from rice paddy fields:
       Summary of available data, Global Biogeochem. Cycles. 19 (2005) 1–10.

[47]   S. Huang, H.K. Pant, J. Lu, Effects of water regimes on nitrous oxide emission
       from soils, Ecol. Eng. 31 (2007) 9–15.

[48]   P. Pramanik, M. Haque, S.Y. Kim, P.J. Kim, C and N accumulations in soil
       aggregates determine nitrous oxide emissions from cover crop treated rice paddy
       soils during fallow season, Sci. Total Environ. 490 (2014) 622–628.
       https://doi.org/10.1016/j.scitotenv.2014.05.046

[49]   D.R. Lovley, The microbe electric : conversion of organic matter to electricity,
       Curr. Opin. Biotechnol. 19 (2008) 564–571.
       https://doi.org/10.1016/j.copbio.2008.10.005

[50]   S. Khan, L. Aijun, S. Zhang, Q. Hu, Y. Zhu, Accumulation of polycyclic aromatic
       hydrocarbons and heavy metals in lettuce grown in the soils contaminated with
       long-term wastewater irrigation, J. Hazard. Mater. 152 (2008) 506–515.
       https://doi.org/10.1016/j.jhazmat.2007.07.014

[51]   C. Abourached, M.J. English, H. Liu, Wastewater treatment by microbial fuel cell
       (MFC) prior irrigation water reuse, J. Clean. Prod. 137 (2016) 144–149.
       https://doi.org/10.1016/j.jclepro.2016.07.048

[52]   K.P.M. Mosse, A.F. Patti, R.J. Smernik, E.W. Christen, T.R. Cavagnaro,
       Physicochemical and microbiological effects of long- and short-term winery
       wastewater application to soils, J. Hazard. Mater. 201–202 (2012) 219–228.
       https://doi.org/10.1016/j.jhazmat.2011.11.071

[53]   Q. Zhou, L. Xu, A. Umar, W. Chen, R. Kumar, Pt nanoparticles decorated SnO_2
       nanoneedles for efficient CO gas sensing applications, Sensors Actuators B.
       Chem. 256 (2018) 656-664. https://doi.org/10.1016/j.snb.2017.09.206

[54]   D. Norton-brandao, S.M. Scherrenberg, J.B. Van Lier, Reclamation of used urban
       waters for irrigation purposes: A review of treatment technologies, J. Environ.
       Manage. 122 (2013) 85–98. https://doi.org/10.1016/j.jenvman.2013.03.012

Enzymatic Fuel Cells   Materials Research Forum LLC
Materials Research Foundations **44** (2019) 131-156   doi: http://dx.doi.org/10.21741/9781644900079-6

Chapter 6

Graphene-based Composites in Enzymatic Biofuel Cells

Mehmet Gülcan[1]*, Gulseren Uzun[2], Buse Guven[2], Fatih Şen[2]*

[1]Department of Chemistry, Faculty of Science, University of Van Yüzüncü Yıl, 65080, Tuşba, Van, Turkey

[2] Sen research Group, Department of Biochemistry, Faculty of Science, University of Dumlupınar, 43100, Kütahya, Turkey

Email: mehmetgulcan65@gmail.com, fatihsen1980@gmail.com

Abstract

Enzymatic biofuel cells [EBFC], are considered to be environmentally friendly energy sources and bio-electrochemical devices. These biofuel cells are operating at ambient temperature and pH. In order to generate electricity, it is necessary to use plant and animal fluids as biofuels in this type of fuel cells. The general working principle is to produce electricity from biological materials as renewable fuels by using the recent technology. These fuel cells are bioelectronic devices that convert chemical energy into electrical energy. This transformation is mostly catalyzed by the use of oxidoreductase enzymes. However, nowadays, graphene and graphene based materials are extremely attractive materials for enzymatic biofuel cells because of their modified surface, high surface areas, electrical, mechanical and thermal properties, and also unique graphitized basal plane structure. This chapter details recent studies on the use of graphene-based composites in the area of enzymatic biofuels.

Keywords

Biofuel Cells, Enzymes, Graphene, Composites, Bioenergy

Contents

1.   Introduction

Fuel cells are electrochemical devices that convert the resulting fuel into electricity and heat energy [1–3]. The power output of a fuel cell is generally expressed by electrochemical reactions occurring in any of the anode and cathode responsible for the reduction of oxidation. As a result, the electrodes must be coated with metallic or metal oxide inorganic electrocatalysts to accelerate the electrochemical reactions [4]. Completion of the cycle is accomplished by a balancing charge over the electrolyte ions [5]. Traditionally, the working principle of fuel cells is to operate with the use of small molecules such as hydrogen or methanol [6–7]. As a new type of fuel cell with a high energy conversion efficiency, biofuel cells can be divided into Enzymatic biofuel cells and microbial fuel cells, which use isolated redox enzymes and living microorganisms as bio-catalysts for power generation, respectively. The different nature of these bio-catalysts leads to various properties and application fields. On the one hand, Enzymatic biofuel cells can produce high levels of power and are well-suited to applications, such as sensors using power outputs as analytical signals, or implantable power sources that are fueled by endogenous substances, such as glucose in the bloodstream. On the other hand, microbial fuel cells offer the possibility of harvesting electricity from a diverse range of 'dirty' fuels that are considered less costly, such as organic waste and organic matter in solids and sediments for wastewater treatment. Despite these exciting possibilities, the expansion of these systems is still limited due to their low power density and poor stability. To further improve the practicality of this technology, a significant focus on understanding the mechanisms of electron transfer and biological processes in biofuel cell systems is required [8].

Materials Research Forum LLC

doi: http://dx.doi.org/10.21741/9781644900079-6

2.   Enzymatic biofuel cell [EBFC]

The Enzymatic biofuel cell was first introduced in 1964 by Yahiro and his colleagues [9], giving rise to many advances in this field, especially in the last decade. The enzymatic biofuel cell works as conventional fuel cells but possesses remarkable advantages over the other fuel cells due to its low cost, renewability, biodegradability, and excellent intrinsic properties, such as specificity towards a substrate and high catalytic activity with a low overvoltage for energy conversion from a substrate, such as glucose in the blood. Recently, the enzymatic biofuel cell has attracted considerable attention because of the possibility of implantation of living batteries into living organisms and electric power generation from the blood sugar in body fluids, which make it a promising power source for cardiac pacemakers, implantable self-powered sensors, and medical devices [8–10–15].

2.1   Basic principles of EBFC

Recent work on enzymatic biofuel cells is about the formation of enzymatic reactions with fuels, such as oxidants, dioxygen, and glucose oxidized in the anode, which are formed by the addition of immobilized enzymes on bioelectrodes, which are reduced in water at the cathode as shown in Fig. 1 [16–18]. The attractive feature of enzymatic biofuel cell is to exploit the excellent selectivity of enzymes, which implies that no other components are required. In this case, both electrodes can be immersed in one chamber containing the substrate and the oxidant, [19–21] in which a membrane is not necessary if there are no crossover reactions between the bioanode and biocathode [22]. As such, the membrane-less enzymatic biofuel cell would be designed for micro- or even nano-power sources, and small enzymatic biofuel cells have great potential for powering implantable medical devices.

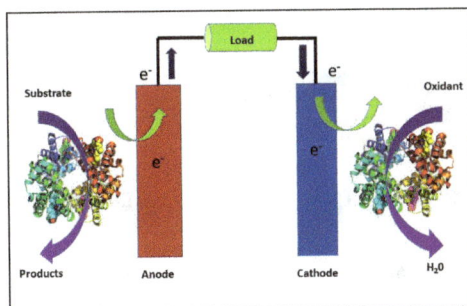

Figure 1. Schematic of a membrane-less Enzymatic biofuel cell

2.2   Electron transfer between enzymes and electrodes

Enzymatic biofuel cells are capable of employing enzymes to catalyze chemical reactions under mild conditions, and they have become eco-friendly and attractive alternative energy sources for microelectronic devices and biosensors. However, because of the electrical insulation of the enzymatic active sites by their surrounding protein shells, the poor electron transfer between enzymes and conductive supports has hindered the applications of enzymatic biofuel cell technology. Thus, a significant fundamental understanding of bio-catalytic enzymes is required, and effective enzyme immobilization methods must be developed to improve the activity and stability of the enzymes. A recent trend in the development of membrane-less enzymatic biofuel cell is the electro-enzymatic oxidation of fuels and oxygen reduction by different enzymes [23–26]. The oxidation of the fuel, typically glucose, occurs at the anode with the aid of glucose oxidase [GOx] or glucose dehydrogenase [GDH] [27–29] while dioxygen is reduced to water at the cathode by specific enzymes, such as laccase or bilirubin oxidase [BOD] [30–33]. These redox reactions may need redox molecules to shuttle electrons from biocatalytic active sites to the electrode surfaces for mediated electron transfer [MET] or undergo direct electron transfer [DET] between the enzyme active sites and the electrode [Fig. 2].

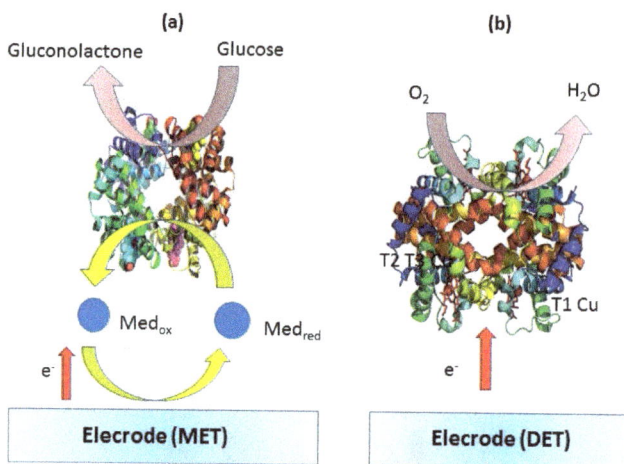

Figure 2. Schematic of the bioelectrocatalytic process at the enzyme bioelectrodes: (a) Mediated electron transfer from redox mediators to a glucose oxidase enzyme on the bioanode surface and (b) Direct electron transfer from a biocathode surface to a laccase enzyme. The essential catalytic copper complexes in the laccase are highlighted

The majority of living organisms have common and abundant amounts of glucose. Therefore, studies on biogases and enzymatic biofuels derived from glucose-oxidase are significant factors in the fact that glucose is a fuel source [34–36]. In glucose oxidase, the redox-active cofactor, flavin adenine dinucleotide [FAD], is buried deep within the protein core making it extremely difficult for direct electron transfer to occur at the electrode [37–39]. To facilitate the electron transfer, small redox mediators that easily diffuse into and out of the enzyme active sites, in particular, those with redox potentials close to that of the flavin adenine dinucleotide cofactor, have been explored to electrically connect these enzymes to the electrode surfaces as shown in Fig. 2a [8–40]. However, this approach has several disadvantages such as high cost, potential toxicity, and limited stability. Because direct electron transfer in the glucose oxidase enzyme depends significantly on the distance between the redox-active cofactor and the electrode surface, several approaches, such as exploring an immobilization strategy for pursuing proper spatial orientation of enzymes to substrates, have been developed to promote the electrical communications of redox enzymes with electrodes. Also, chemically modified electrodes with favourable functional materials, especially nanostructured materials, could display electron-mediating functions and facilitate the direct electron transfer processes of redox enzymes. Many researchers have worked on engineering and high-performance bioanodes. Along with these studies, considerable progress has been made on glucose oxidase-based enzymatic biofuels. Also, enzymatic biofuels cells can be successfully associated with living organisms, portable electronic devices operating with sugar, sensors and potential applications that can be practically adapted [41–43].

To date, laccase and bilirubin oxidase have been utilized at the biocathode to perform an efficient oxygen reduction reaction [ORR], mainly relying on the high catalytic capabilities of these two enzymes with coordinated copper [Cu] centers to catalyze the reduction of oxygen to water [44–47]. As a multicopper oxidase [MCO], laccase is one of the most important and widely used enzymes for cathodic oxygen reduction reaction because the redox potential of the active site of laccase [~0.78 V vs. SHE at neutral pH] is very close to that of the O_2/H_2O redox couple. The active center of laccase contains four Cu atoms located in three different regions of the protein which are classified as T1, T2 and T3 [Fig. 2b]. The T1-Cu site is a hydrophobic pocket near the surface of the protein and can accept electrons from substrates that act as electron donors with the electrode. Then, electrons are further passed to the T2/T3 Cu sites formed by a trinuclear cluster where oxygen is efficiently reduced to water in a four-electron mechanism. Aside from laccase, several bilirubin oxidase biocathodes have also been established for oxygen reduction reaction catalysis with similar catalytic mechanisms. Unlike the optimal pH of laccase [B5], bilirubin oxidase operates optimally at pH 7. Also, the redox potential of

Materials Research Forum LLC
doi: http://dx.doi.org/10.21741/9781644900079-6

laccase T1-Cu is higher than that of bilirubin oxidase; hence, its activity is higher than that of the commercial bilirubin oxidase. Laccase is widely used as the cathode catalyst. Because the T1-Cu site in the enzyme is responsible for the electron transfer between the substrate and enzyme, the direct electron transfer of laccase at the electrode surface can be governed by the inherent redox potential of the enzyme active sites and suitable enzyme immobilization with nanostructured materials to ensure the electrode materials are sufficiently close to the T1-Cu site [8].

2.3 Graphene and its characteristic features

Graphite is a two-dimensional layer with sp2 hybridized carbon atoms which has excellent electrical conductivity. It is also considered an electrode material for the extremely high specific surface area [up to 2600 m^2/g] and electrochemical devices providing new possibilities for biofuel cells [19–22]. The overall preparation of the graphene is the result of the reduction of the graphene oxide. Compared with reduced graphene oxide and graphene, the oxidant is more hydrophilic regarding the presence of reduced oxygen-containing functional groups [48]. Improved hydrophilicity benefits the immobilization of biocatalysts. Graphene production can be obtained in many different ways, as presented in Figures 3 and 4. Also, in many articles, the synthesis of the graphene production is explained in detail, so there will be no further discussion regarding its production here. Although a pure graphene production benefits from the understanding of the graphene basis, it is not preferred for electrochemical sensors because of its high cost and inability to scale in production [49].

Hence, other forms of graphene were investigated for several reasons. As summarized in Figure 4, reduced graphene oxide is formed by oxidation/degradation of the graphene and reduction by various means, such as chemical, thermal and electrochemical [49]. Furthermore, reduced graphene oxide offers plenty of space for chemical functionalization which would improve the interaction between reduced graphene oxide and biocatalysts. The improved hydrophilicity and interaction would play an essential role in the direct electron transfer. At another level, graphene can be given a new porous architectural feature through its various physical properties and unique structure. Thanks to this feature gained in graphene, active electrochemical surfaces can be obtained for collection of electrons as well as ease of mass transfer of mediators. This can be preferred because the graphite increases the efficiency of biocatalyst and electron transfer as electrode materials [8].

Enzymatic Fuel Cells   Materials Research Forum LLC
Materials Research Foundations **44** (2019) 131-156   doi: http://dx.doi.org/10.21741/9781644900079-6

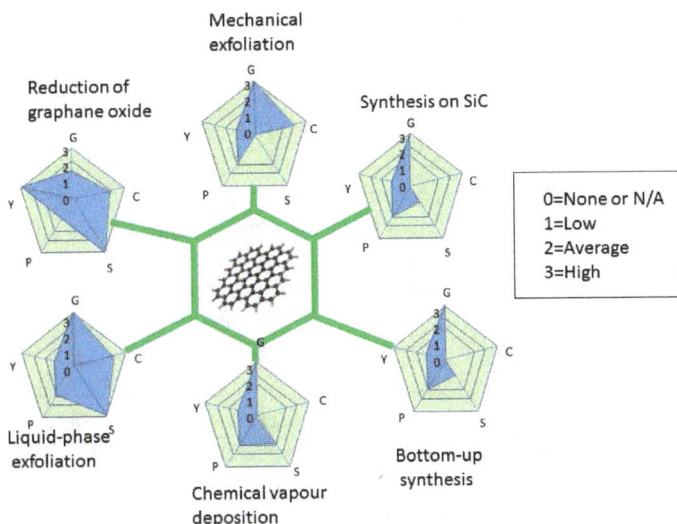

Figure 3. The most common graphene production methods. Each method has been evaluated regarding graphene quality (G), cost aspect (C; a low value corresponds to the high cost of production), scalability (S), purity (P) and yield (Y) of the overall production process

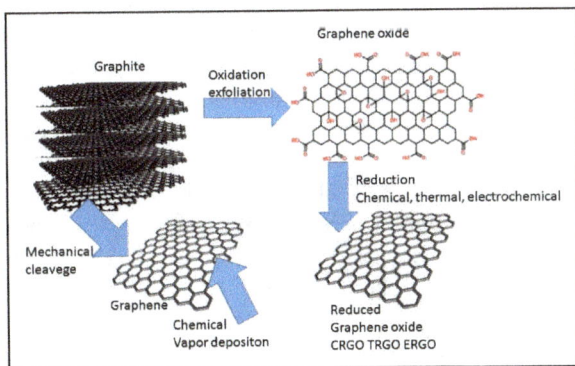

Figure 4. A schematic illustration of possible ways for the preparation of graphene and reduced graphene oxide

Materials Research Forum LLC
doi: http://dx.doi.org/10.21741/9781644900079-6

2.4 Graphene as an electrode material for enzymatic biofuel cells

The production of enzymatic bioelectrodes is accomplished by depositing graphene [bare, functional, or composite] enzymes on the surface electrodes [50–52]. Due to the providence and montage methods, the properties of the graphene layers can be changed in large quantities. The performance of bioelectrodes can be dictated by the physicochemical properties of the graphene and deposited layers. The first method to look to separate graphene plates from pyrolytic graphite is mechanical peeling [53–54]. In addition to the mechanical peeling method, there are chemical vapour units, epitaxial growth, electrical arc discharge, non-compaction of carbon nanotubes, and reduction of gross surface oxides [graphene oxide] [54–56]. The most important of these is chemical vapour deposition because of its ability to manufacture wide field and functionalized graphene layers [57–58]. Providence conditions, sizes, number of sheet, and constitutive imperfections on graphene properties are quite effective [59]. Highly defective structures cause different and unexpected features of the graphene. Graphene has semi-metallic and metallic conductive features in single-layer [SLG] and multilayer [MLG, more than 10 layers] states [60]. As changes in the number of layers show different conducting properties, enzyme conjugation does not change [61–62]. Although graphite is a promising nonmetallic material, it is very difficult to use it directly in technological studies. There are some situations to be considered for the use of graphene in practice, such as graphene imperfections, edge configuration types, reactivity, low resolution, and preparation methods [63]. The element that prevents the direct use of the graphene is the π-electrons in the electronic construction of the insulated graphene. Surface modification of the graphene can provide useful properties through functional groups [64]. Metallic contaminants in the landscape affect the electrochemical properties of regulating devices [65]. Due to the entropic effect, it is quite difficult to produce perfect graphene sheets [63–66]. Some imperfections can come into play automatically. Significant changes in the properties of the graphene occur due to the atoms forming the defects in the graphene structure [67–68]. Graphene can also show the strong covalent bonds throughout the structure as well as weak van der Waals bonds between layers. The extremely high surface area of the graphene and the weakness of the van der Waals bond on the plate prevent it from becoming single-layer graphene [54]. The tendency to form irreversible aggregates occurs between the graphene structure [69].

As a result, intact graphene does not exhibit significant solubility characteristics in many solvents. Increasing the solubility is achieved by forming hydrophilic and hydrophilic groups on the graphene structure. The newly formed bonds contribute to changes in physical and chemical properties and changes in electrochemical behaviour in conductivity and functionality. The changes that occur in the structure are significant to link the enzyme which affects the performance of the resulting electrons. Most of the bioelectrochemical applications contain functionalized graphene [62]. There are three primary ways to change the surface of graphene. These are covalent interactions [70–71], non-covalent interactions [72–73], and nanoparticles [NPs] [73]. The covalent attachment state is due to a chemical transformation by covalent addition of sp^2 carbon to sp^3 hybridization [74]. The non-covalent functionalization polymer coating method is associated with absorptive biomolecules or small aromatic molecules on the graphene surface. This association can be achieved by absorption of surface-active substances and by physical coactions such as hydrophobic, electrostatic or van der Waals forces [62].

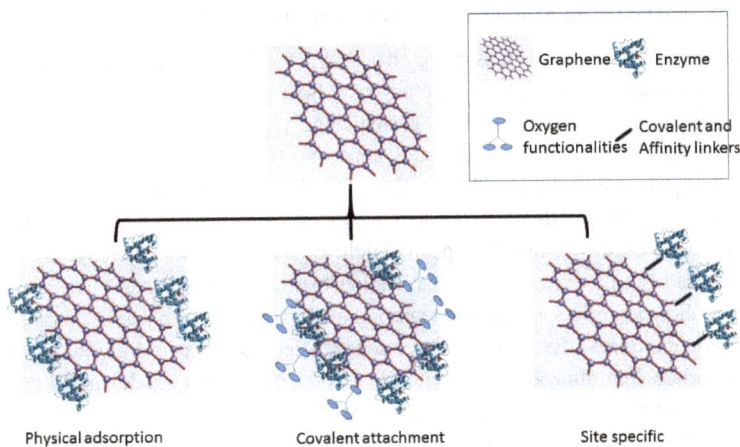

Figure 5. Demonstration of enzyme immobilization methods on graphene species

2.5 Immobilization of enzymes onto graphene derivatives

There are two important parameters in terms of the efficiency of enzymatic electrons: the first one is the redox link between the enzyme and the electrode, and the second one is the enzyme activity. In Enzymatic biofuel cell, the enzyme is fixed around the electrode which is in charge of the anode or cathode. The methods and materials that enable the enzyme to remain on the electrode when immobilized are very important. Among this method and materials, the electrical connection is used to adjust the critical states of enzymes. The situation that facilitates easy binding of enzymes is related to the presence or absence of functional groups on the graphene surface. Physical absorption, covalent attachment, trapping, self-assembly, and affinity are among the traditional strategies. The application of these strategies is achieved by the interaction of enzymes on the functionalized graphene [75]. The biological activity of the immobilized enzyme and the efficiency of the electron transfer depend on the interaction between the enzyme and the graphene material. Fig. 5 summarizes the various strategies that can be used to immobilize enzymes onto graphene through physical adsorption, covalent attachment, and site-specific affinity interactions. Also, direct immobilization on graphene and customized graphene can combine nanocomposite materials, graphene, and other materials [62–76–77].

2.6   Graphene-based composites in enzymatic biofuel cells

Enzyme-based biofuels known as energy harvesting methods from organic materials have attracted considerable interest in recent years. It is attractive as a sustainable energy source with bioelectronics, which is a small and pluggable feature [78]. At present, studies on specific density bioelectrodes continue to improve performance, especially concerning power density [5–78–81]. Essential developments in this area have been achieved by exploiting the superior properties of carbon-based nanomaterials. Those materials were used to expand a specific field of the electrode. Thereby, considerable progress has been observed in the area leading to an improved current density [82–83]. Thanks to these materials, efficient integration make it easy to come across biological and electronic components. Expansion of electron transfer and enzyme stability is facilitated by appropriate immobilization techniques. Efforts have been made to develop these techniques that will lead to the formation of efficient nanocatalysts for high-performance enzymatic biofuel cells [84–86]. As noted earlier, the new methods for biomolecule immobilization are possible thanks to graphene-based nanostructures. This enabled the production of small and high-performance enzymatic biofuel cells [86–88]. In the first reported graphene-based biofuel cell, a sol-gel matrix and graphene holders were used, and the enzyme [glucose oxidase and bilirubin oxidase [bilirubin oxidase], redox mediators, anodes and cathodes were coordinated [89]. In another report, in which a polypyrrole conductive polymer is observed to grow electrochemically in polypyrrole conducting polymer, glucose oxidase was applied to produce an electrode by capturing electrochemical growth [90–94]. This method facilitates the binding of high amounts of the enzyme to the porous structure of polyproline with a 3-D configuration. Power output enzyme stability is better than that of glucose oxidase iodine which blocks the enzyme and graphene in a sol-gel plate [95]. A 3-D structure and improved performance of the conductive polypropylene film providing improved electron transfer has resulted in a higher diffusion rate [96–98]. Lee et al. [96] produced a glucose oxidase/laccase biofuels cell using the kit site and cobalt hydroxide hybrid composite together with graphene oxide. The biofuel cell produced showed high electrochemical features and power density. Besides, Devadas et al. [99] used carbon nanotube and reduced graphene oxide modified based electrodes as electrode materials. The active surface area related to the properties of the nanomaterials to obtain a synergistic effect [100–104]. The electrocatalytic features were obtained by inhibiting O_2 reduction. Power was generated at a density of 46 µW/cm^2. Graphene oxide is the material which is similar to the graphene shape used in bioelectrocatalysis, but depending on the reduction pathway, the change in the current density of the enzymatic biofuel cells based on these materials is observed and coupled to the electrode surface [105]. In a sample, the current density of the resulting biocation was measured after immobilization of bilirubin oxidase on electrochemically reduced graphene oxide. In the first measurements, the electrochemical reduction of graphene oxide with bilirubin oxidase enhanced the obtained values. Also, the small-sized graphene oxide layers and the purified graphene oxide are incubated [106]. However, a maximum flow density of 57.8 µW / cm^2 was achieved by a glucose oxidase bioanode and laccase biocathode with enzymes that block graphene nanosheets by cross-linking glutaraldehyde [82]. Additional binders need to be used to deposit many protocol enzymes and graphene materials on the electrode surfaces. This is the process without the need for any binder or crosslinker [107]. While the process proceeds in the form of a graphene oxide-glucose oxidase composite on the glassy carbon electrode, the reduced graphene oxide is electrochemically reduced by a continuous potential cycle between 0 and 1.5 V. The biocathode is formed by blocking laccase in multiwall carbon nanotubes [MWCNT]/zinc oxide [ZnO] composite [108].

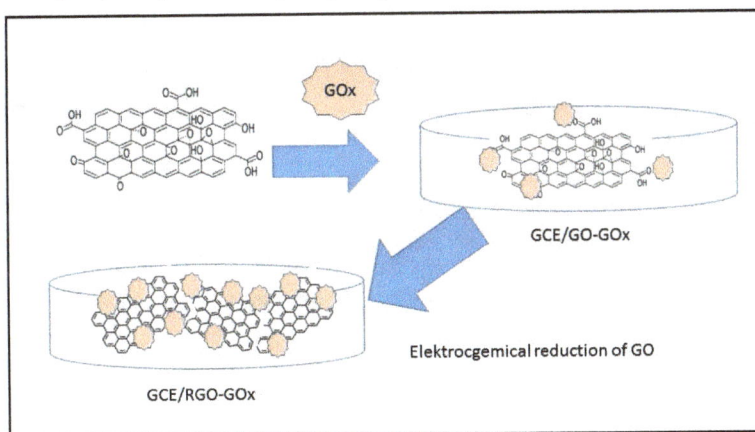

Figure 6. Single-stage production of graphene oxide-glucose oxidase biocomposite

In studies on enzymatic biofuel cells performed recently, 3D-graft-based nanostructure configurations at advanced dimensions are being explored to enhance the active surface area of the materials [59–109–111]. Because of its large surface area, the 3D structure allows for electron transfer and high enzyme loading. In the 3D-graphene-Single-walled Carbon Nanotubes - glucose oxidase-hybrid electron voltammetry analysis with 1 mM glucose, despite the presence of direct electron transfer in the biogas, much more negative potential glucose oxidation was observed compared to the glucose oxidase-I redox potential [112–113]. Figure 7 shows the configurations of glucose oxidase/ Laccase graphene- Single-walled Carbon Nanotubes hybrid bioelectrodes. Zhang et al. reported that [110], reduction of bioelectrocatalytic oxygen with the help of laccase was blocked using 3D graphene nets as a new substrate [110].

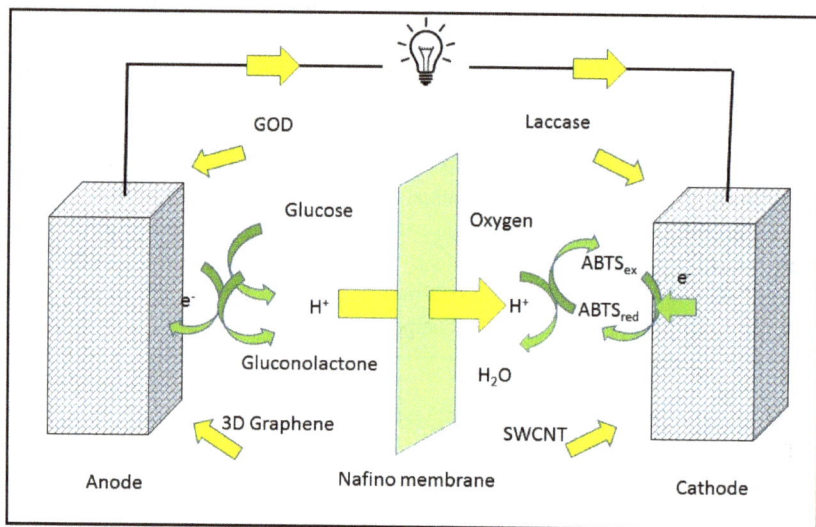

Figure 7. Illustration of the Enzymatic biofuel cell based on 3D graphene-Single-walled Carbon Nanotubes hybrid electrodes

Materials Research Forum LLC

doi: http://dx.doi.org/10.21741/9781644900079-6

Table 1. The characteristic features of developed graphene based-enzymatic biofuel cells.

Biofuel cell [BFC]	Graphene-based material	OCP [mV]	pMAX [$\mu W/cm^2$]	pMAX [mV]	Ref.
TMOS gel + CRGO + FM + GOx on Au plate TMOS gel + CRGO + ABTS + BOD on Au	CRGO	580	24.3	380	97
Au plate/CRGO/FM+GOx/PPy Au plate/CRGO/ABTS+Lac/PPy	CRGO	790	78.3	500	98
GCE/graphene nanoplates/GOx/Nafion GCE/[GR]/Lac+BSA/Glutaraldehyde	Nanoscale graphene platelet	550	58.0	220	115
GCE/ERGO-MWCNTs/GOx/Nafion GCE/CRGO + PtNPs +Nafion	ERGO/CRGO	400	46	80	107
GCE/ER [GO + GOx] GCE/MWCNT-ZnO/Lac	ERGO	60	0.054	50	117
AuE/electrodeposited [GO+Co[OH]$_2$ in CHI] + GOx AuE/electrodeposited [GO/Co[OH]$_2$/CHI]/Lac	ERGO	600	517	460	104
G/CNTs-COOH/GOx G/CNTs-COOH/Lac + ABTS in solution	3D graphene decorated with SWCNTs	1200	2270	500	121
Nafion/GOx/ferrocene/3D-GNs/GCE Nafion/Lac/3D-GNs-PTCA-DA/GCE	3D graphene network	960	112	190	120
GCE/graphene/MWCNTs/MG/GDH GCE/SWCNTs/[Lac+BSA+Glutaraldehyde]	Graphene	690	22.50	480	126
PVA-SbQ/*Ct*CDH/*Th*Lac AuNPs/GPHbased	Modified graphene	-	2.15	-	127
PVA-SbQ/ *Ct*CDH/AuNPs/GPH‖ *Th*Lac/AuNPs/G	Modified graphene	-	1.57	-	127
GCE/GR-CoPc/GOx GCE/GR-FePc	Graphene-cobalt phthalocyanine	-	23	-	128

ERGO-electrochemically reduced graphene oxide; Lac–laccase; GPH-graphene-based screen-printed electrode; PTCA-3,4,9,10-perylene tetracarboxylic acid; GOx-Glucose oxidase; DA-dopamine; BOD–bilirubin oxidase; FM-ferrocenemethanol; FMCA–ferrocene monocarboxylic acid; CHI-chitosan; GDH- glucose dehydrogenase, GR-CoPc-graphene-cobalt phthalocyanine; GR-FePc- graphene-iron phthalocyanine; G-graphite based screen-printed electrode; PPy–polypyrrole [114]; MG-methylene green [115]; PVA-SbQ-poly[vinyl alcohol] N-methyl-4[4'-formylstyryl]pyridinium methosulfate acetal, ThLac-Trametes hirsuta laccase; BSA-bovine serum albumin; TMOS-tetramethoxysilane; GCE-glassy carbon electrode; CRGO-chemically reduced graphene oxide ABTS-2,2'-azino-bis[3-ethyl benzothiazole-6-sulphonic acid are also recently used materials in literature.

Conclusions

As a conclusion, enzymatic biofuel cells [EBFC] are bioelectronic devices that convert chemical energy into electrical energy. The general working principle is to produce electricity from biological materials as renewable fuels by using some enzymes and/or carbon-based materials. Especially, graphene and graphene-based materials are extremely attractive materials for enzymatic biofuel cells because of their modified surface, high surface areas, electrical, mechanical and thermal properties, and also unique graphitized basal plane structure. This chapter focuses mostly on recent studies on the use of graphene-based composites in the area of enzymatic biofuels.

References

[1]   R. A. S. Luz, A. R. Pereira, J. C. P. de Souza, F. C. P. F. Sales, & F. N. Crespilho, Enzyme Biofuel Cells: Thermodynamics, Kinetics and Challenges in Applicability. ChemElectroChem, 1 (2014) 1751–1777. https://doi.org/10.1002/celc.201402141.

[2]   W. R. Grove, XXIV. On Voltaic Series and The Combination of Gases by Platinum. The London, Edinburgh, and Dublin Philosophical Magazine and Journal of Science, 14 (1839) 127–130. https://doi.org/10.1080/14786443908649684.

[3]   C. F. Schœnbein, X. On The Voltaic Polarization of Certain Solid and Fluid Substances. Philosophical Magazine Series 3, 14 (1839) 43–45. https://doi.org/10.1080/14786443908649658.

[4]   J. A. Cracknell, K. A. Vincent, & F. A. Armstrong, Enzymes as Working or Inspirational Electrocatalysts for Fuel Cells and Electrolysis. Chemical Reviews, 108 (2008) 2439–2461. https://doi.org/10.1021/cr0680639.

Materials Research Forum LLC
doi: http://dx.doi.org/10.21741/9781644900079-6

[5]   R. A. Bullen, T. C. Arnot, J. B. Lakeman, & F. C. Walsh, Biofuel Cells and Their
       Development. Biosensors and Bioelectronics, 21 (2006) 2015–2045.
       https://doi.org/10.1016/j.bios.2006.01.030.

[6]   L. Carrette, K. A. Friedrich, & U. Stimming, Fuel Cells: Principles, Types, Fuels,
       and Applications. ChemPhysChem, 1 (2000) 162–193.
       https://doi.org/10.1002/1439-7641(20001215)1:4<162::AID-CPHC162>3.0.CO;2-
       Z.

[7]   A. Mitsos, I. Palou-Rivera, & P. I. Barton, Alternatives for Micropower
       Generation Processes. Industrial & Engineering Chemistry Research, 43 (2004)
       74–84. https://doi.org/Doi 10.1021/Ie0304917.

[8]   C. E. Zhao, P. Gai, R. Song, Y. Chen, J. Zhang, & J. J. Zhu, Nanostructured
       Material-Based Biofuel Cells: Recent Advances and Future Prospects. Chemical
       Society Reviews, 46 (2017) 1545–1564. https://doi.org/10.1039/c6cs00044d.

[9]   A. T. Yahiro, S. M. Lee, & D. O. Kimble, Bioelectrochemistry. I. Enzyme
       Utilizing Bio-Fuel Cell Studies. BBA - Specialised Section On Biophysical
       Subjects, 88 (1964) 375–383. https://doi.org/10.1016/0926-6577(64)90192-5.

[10]   M. Gamella, N. Guz, S. Mailloux, J. M. Pingarrón, & E. Katz, Activation of A
       Biocatalytic Electrode by Removing Glucose Oxidase from The Surface-
       Application to Signal Triggered Drug Release. ACS Applied Materials and
       Interfaces, 6 (2014) 13349–13354. https://doi.org/10.1021/am504561d.

[11]   M. Rasmussen & S. D. Minteer, Self-Powered Herbicide Biosensor Utilizing
       Thylakoid Membranes. Analytical Methods, 5 (2013) 1140–1144.
       https://doi.org/10.1039/c3ay26488b.

[12]   X. Zhang, L. Zhang, Q. Zhai, W. Gu, J. Li, & E. Wang, Self-Powered Bipolar
       Electrochromic Electrode Arrays for Direct Displaying Applications. Analytical
       Chemistry, 88 (2016) 2543–2547. https://doi.org/10.1021/acs.analchem.6b00054.

[13]   K. MacVittie, T. Conlon, & E. Katz, A Wireless Transmission System Powered by
       An Enzyme Biofuel Cell Implanted in An Orange. Bioelectrochemistry, 106
       (2015) 28–33. https://doi.org/10.1016/j.bioelechem.2014.10.005.

[14]   A. A. Babadi, S. Bagheri, & S. B. A. Hamid, Progress on Implantable Biofuel
       Cell: Nano-Carbon Functionalization for Enzyme Immobilization Enhancement.
       Biosensors and Bioelectronics, 79 (2016) 850–860.
       https://doi.org/10.1016/j.bios.2016.01.016.

[15]   U. Schröder, From In vitro to In vivo-biofuel Cells are Maturing. Angewandte
       Chemie - International Edition, 51 (2012) 7370–7372.
       https://doi.org/10.1002/anie.201203259.

[16]   X.-Y. Yang, G. Tian, N. Jiang, & B.-L. Su, Immobilization Technology: A
       Sustainable Solution for Biofuel Cell Design. Energy Environ. Sci., 5 (2012)
       5540–5563. https://doi.org/10.1039/C1EE02391H.

[17]   L. Deng, F. Wang, H. Chen, L. Shang, L. Wang, T. Wang, & S. Dong, A Biofuel
       Cell with Enhanced Performance by Multilayer Biocatalyst Immobilized on
       Highly Ordered Macroporous Electrode. Biosensors and Bioelectronics, 24 (2008)
       329–333. https://doi.org/10.1016/j.bios.2008.04.006.

[18]   C. Hou, D. Yang, B. Liang, & A. Liu, Enhanced Performance of A Glucose/O2
       Biofuel Cell Assembled with Laccase-Covalently Immobilized Three-Dimensional
       Macroporous Gold Film-Based Biocathode and Bacterial Surface Displayed
       Glucose Dehydrogenase-Based Bioanode. Analytical Chemistry, 86 (2014) 6057–
       6063. https://doi.org/10.1021/ac501203n.

[19]   A. K. Geim & K. S. Novoselov, The Rise of Graphene. Nature Materials, 6 (2007)
       183–191. https://doi.org/10.1038/nmat1849.

[20]   W. Gao, The Chemistry of Graphene Oxide. Graphene Oxide: Reduction Recipes,
       Spectroscopy, and Applications, (2015) 61–95. https://doi.org/10.1007/978-3-319-
       15500-5_3.

[21]   M. J. Allen, V. C. Tung, & R. B. Kaner, Honeycomb Carbon: A Review of
       Graphene. Chemical Reviews, 110 (2010) 132–145.
       https://doi.org/10.1021/cr900070d.

[22]   D. Li, M. B. Müller, S. Gilje, R. B. Kaner, & G. G. Wallace, Processable Aqueous
       Dispersions of Graphene Nanosheets. Nature Nanotechnology, 3 (2008) 101–105.
       https://doi.org/10.1038/nnano.2007.451.

[23]   R. L. Arechederra, B. L. Treu, & S. D. Minteer, Development of Glycerol/O2
       Biofuel Cell. Journal of Power Sources, 173 (2007) 156–161.
       https://doi.org/10.1016/j.jpowsour.2007.08.012.

[24]   C. M. Moore, S. B. Minteer, & R. S. Martin, Microchip-Based Ethanol/Oxygen
       Biofuel Cell. Lab on a Chip, 5 (2005) 218–225. https://doi.org/10.1039/b412719f.

[25]   M. J. Moehlenbrock, T. K. Toby, A. Waheed, & S. D. Minteer, Metabolon
       Catalyzed Pyruvate/Air Biofuel Cell. Journal of the American Chemical Society,
       132 (2010) 6288–6289. https://doi.org/10.1021/ja101326b.

[26]   S. Topcagic & S. D. Minteer, Development of A Membraneless Ethanol/Oxygen
       Biofuel Cell. Electrochimica Acta, 51 (2006) 2168–2172.
       https://doi.org/10.1016/j.electacta.2005.03.090.

Materials Research Forum LLC
doi: http://dx.doi.org/10.21741/9781644900079-6

[27]   E. Katz & I. Willner, A Biofuel Cell with Electrochemically Switchable and Tunable Power Output. Journal of the American Chemical Society, 125 (2003) 6803–6813. https://doi.org/10.1021/ja034008v.

[28]   H. B. Noh, M. S. Won, J. Hwang, N. H. Kwon, S. C. Shin, & Y. B. Shim, Conjugated Polymers and An Iron Complex as Electrocatalytic Materials for An Enzyme-Based Biofuel Cell. Biosensors and Bioelectronics, 25 (2010) 1735–1741. https://doi.org/10.1016/j.bios.2009.12.020.

[29]   Y. Yan, W. Zheng, L. Su, & L. Mao, Carbon-Nanotube-Based Glucose/O2 Biofuel Cells. Advanced Materials, 18 (2006) 2639–2643. https://doi.org/10.1002/adma.200600028.

[30]   M. Ammam & J. Fransaer, Combination of Laccase and Catalase in Construction of H2O2-O2 Based Biocathode for Applications in Glucose Biofuel Cells. Biosensors and Bioelectronics, 39 (2013) 274–281. https://doi.org/10.1016/j.bios.2012.07.066.

[31]   J. Filip, J. Šefčovičová, P. Gemeiner, & J. Tkac, Electrochemistry of Bilirubin Oxidase And Its Use in Preparation of A Low Cost Enzymatic Biofuel Cell Based on A Renewable Composite Binder Chitosan. Electrochimica Acta, 87 (2013) 366–374. https://doi.org/10.1016/j.electacta.2012.09.054.

[32]   A. Habrioux, T. Napporn, K. Servat, S. Tingry, & K. B. Kokoh, Electrochemical Characterization of Adsorbed Bilirubin Oxidase on Vulcan XC 72R For The Biocathode Preparation in A Glucose/O2biofuel Cell. Electrochimica Acta, 55 (2010) 7701–7705. https://doi.org/10.1016/j.electacta.2009.09.080.

[33]   G. Gupta, C. Lau, B. Branch, V. Rajendran, D. Ivnitski, & P. Atanassov, Direct Bio-Electrocatalysis By Multi-Copper Oxidases: Gas-Diffusion Laccase-Catalyzed Cathodes for Biofuel Cells. Electrochimica Acta, 56 (2011) 10767–10771. https://doi.org/10.1016/j.electacta.2011.01.089.

[34]   F. Gao, Y. Yan, L. Su, L. Wang, & L. Mao, An Enzymatic Glucose/O2 Biofuel Cell: Preparation, Characterization and Performance In Serum. Electrochemistry Communications, 9 (2007) 989–996. https://doi.org/10.1016/j.elecom.2006.12.008.

[35]   D. Ivnitski, B. Branch, P. Atanassov, & C. Apblett, Glucose Oxidase Anode for Biofuel Cell Based on Direct Electron Transfer. Electrochemistry Communications, 8 (2006) 1204–1210. https://doi.org/10.1016/j.elecom.2006.05.024.

[36]   A. Zebda, C. Gondran, P. Cinquin, & S. Cosnier, Glucose Biofuel Cell Construction Based on Enzyme, Graphite Particle and Redox Mediator

Compression. Sensors and Actuators, B: Chemical, 173 (2012) 760–764. https://doi.org/10.1016/j.snb.2012.07.089.

[37] J. T. Holland, C. Lau, S. Brozik, P. Atanassov, & S. Banta, Engineering of Glucose Oxidase for Direct Electron Transfer via Site-Specific Gold Nanoparticle Conjugation. Journal of the American Chemical Society, 133 (2011) 19262–19265. https://doi.org/10.1021/ja2071237.

[38] C. Schulz, R. Kittl, R. Ludwig, & L. Gorton, Direct Electron Transfer from The FAD Cofactor of Cellobiose Dehydrogenase to Electrodes. ACS Catalysis, 6 (2016) 555–563. https://doi.org/10.1021/acscatal.5b01854.

[39] Y. Yamashita, S. Ferri, M. L. Huynh, H. Shimizu, H. Yamaoka, & K. Sode, Direct Electron Transfer Type Disposable Sensor Strip for Glucose Sensing Employing An Engineered FAD Glucose Dehydrogenase. Enzyme and Microbial Technology, 52 (2013) 123–128. https://doi.org/10.1016/j.enzmictec.2012.11.002.

[40] F. Barrière, Y. Ferry, D. Rochefort, & D. Leech, Targetting Redox Polymers as Mediators for Laccase Oxygen Reduction in A Membrane-Less Biofuel Cell. Electrochemistry Communications, 6 (2004) 237–241. https://doi.org/10.1016/j.elecom.2003.12.006.

[41] A. Zebda, S. Cosnier, J. P. Alcaraz, M. Holzinger, A. Le Goff, C. Gondran, F. Boucher, F. Giroud, K. Gorgy, H. Lamraoui, & P. Cinquin, Single Glucose Biofuel Cells Implanted in Rats Power Electronic Devices. Scientific Reports, 3 (2013) 1–5. https://doi.org/10.1038/srep01516.

[42] L. Halámková, J. Halámek, V. Bocharova, A. Szczupak, L. Alfonta, & E. Katz, Implanted Biofuel Cell Operating in A Living Snail. Journal of the American Chemical Society, 134 (2012) 5040–5043. https://doi.org/10.1021/ja211714w.

[43] K. Shoji, Y. Akiyama, M. Suzuki, N. Nakamura, H. Ohno, & K. Morishima, Biofuel Cell Backpacked Insect and Its Application to Wireless Sensing. Biosensors and Bioelectronics, 78 (2016) 390–395. https://doi.org/10.1016/j.bios.2015.11.077.

[44] S. Garavaglia, M. T. Cambria, M. Miglio, S. Ragusa, V. Iacobazzi, F. Palmieri, C. D'Ambrosio, A. Scaloni, & M. Rizzi, The Structure of Rigidoporus Lignosus Laccase Containing A Full Complement of Copper Ions, Reveals An Asymmetrical Arrangement for The T3 Copper Pair. Journal of Molecular Biology, 342 (2004) 1519–1531. https://doi.org/10.1016/j.jmb.2004.07.100.

[45] A. Christenson, S. Shleev, N. Mano, A. Heller, & L. Gorton, Redox Potentials of The Blue Copper Sites of Bilirubin Oxidases. Biochimica et Biophysica Acta -

Bioenergetics, 1757 (2006) 1634–1641.
https://doi.org/10.1016/j.bbabio.2006.08.008.

[46] V. Soukharev, N. Mano, & A. Heller, A Four-Electron O2-Electroreduction Biocatalyst Superior to Platinum and a Biofuel Cell Operating at 0.88 V. Journal of the American Chemical Society, 126 (2004) 8368–8369. https://doi.org/10.1021/ja0475510.

[47] J. Cline, B. Reinhammar, P. Jensen, R. Venters, & B. M. Hoffman, Coordination Environment for The Type 3 Copper Center of Tree Laccase And Cub of Cytochrome C Oxidase as Determined by Electron Nuclear Double Resonance. Journal of Biological Chemistry, 258 (1983) 5124–5128.

[48] V. Georgakilas, J. N. Tiwari, K. C. Kemp, J. A. Perman, A. B. Bourlinos, K. S. Kim, & R. Zboril, Noncovalent Functionalization of Graphene and Graphene Oxide for Energy Materials, Biosensing, Catalytic, and Biomedical Applications. Chemical Reviews, 116 (2016) 5464–5519. https://doi.org/10.1021/acs.chemrev.5b00620.

[49] S. J. Rowley-Neale, E. P. Randviir, A. S. Abo Dena, & C. E. Banks, An Overview of Recent Applications of Reduced Graphene Oxide as A Basis of Electroanalytical Sensing Platforms. Applied Materials Today, 10 (2018) 218–226. https://doi.org/10.1016/j.apmt.2017.11.010.

[50] B. Sen, B. Demirkan, A. Şavk, S. Karahan Gülbay, & F. Sen, Trimetallic PdRuNi Nanocomposites Decorated on Graphene Oxide: A Superior Catalyst for The Hydrogen Evolution Reaction. International Journal of Hydrogen Energy, 43 (2018) 17984–17992. https://doi.org/10.1016/j.ijhydene.2018.07.122.

[51] S. Eris, Z. Daşdelen, Y. Yıldız, & F. Sen, Nanostructured Polyaniline-rGO Decorated Platinum Catalyst with Enhanced Activity and Durability for Methanol Oxidation. International Journal of Hydrogen Energy, 43 (2018) 1337–1343. https://doi.org/10.1016/j.ijhydene.2017.11.051.

[52] S. Eris, Z. Daşdelen, & F. Sen, Investigation of Electrocatalytic Activity and Stability Of Pt@F-VC Catalyst Prepared by In-Situ Synthesis for Methanol Electrooxidation. International Journal of Hydrogen Energy, 43 (2018) 385–390. https://doi.org/10.1016/j.ijhydene.2017.11.063.

[53] K. S. Novoselov, Electric Field Effect in Atomically Thin Carbon Films. Science, 306 (2004) 666–669. https://doi.org/10.1126/science.1102896.

[54] T. Kuila, S. Bose, A. K. Mishra, P. Khanra, N. H. Kim, & J. H. Lee, Chemical Functionalization of Graphene and Its Applications. Progress in Materials Science, 57 (2012) 1061–1105. https://doi.org/10.1016/j.pmatsci.2012.03.002.

[55] C. Soldano, A. Mahmood, & E. Dujardin, Production, Properties and Potential of Graphene. Carbon, 48 (2010) 2127–2150. https://doi.org/10.1016/j.carbon.2010.01.058.

[56] Y. Zhu, S. Murali, W. Cai, X. Li, J. W. Suk, J. R. Potts, & R. S. Ruoff, Graphene and Graphene Oxide: Synthesis, Properties, and Applications. Advanced Materials, 22 (2010) 3906–3924. https://doi.org/10.1002/adma.201001068.

[57] M. L. T. Cossio, L. F. Giesen, G. Araya, M. L. S. Pérez-Cotapos, R. L. VERGARA, M. Manca, R. A. Tohme, S. D. Holmberg, T. Bressmann, D. R. Lirio, J. S. Román, R. G. Solís, S. Thakur, S. N. Rao, E. L. Modelado, A. D. E. La, C. Durante, U. N. A. Tradición, M. En, E. L. Espejo, D. E. L. A. S. Fuentes, U. A. De Yucatán, C. M. Lenin, L. F. Cian, M. J. Douglas, L. Plata, & F. Héritier, Large-Area Synthesis of High-Quality and Uniform Graphene Films on Copper Foils. Science Mag, XXXIII (2012) 81–87. https://doi.org/10.1007/s13398-014-0173-7.2.

[58] K. S. Kim, Y. Zhao, H. Jang, S. Y. Lee, J. M. Kim, K. S. Kim, J. H. Ahn, P. Kim, J. Y. Choi, & B. H. Hong, Large-Scale Pattern Growth of Graphene Films for Stretchable Transparent Electrodes. Nature, 457 (2009) 706–710. https://doi.org/10.1038/nature07719.

[59] S. Nardecchia, D. Carriazo, M. L. Ferrer, M. C. Gutiérrez, & F. Del Monte, Three Dimensional Macroporous Architectures and Aerogels Built of Carbon Nanotubes and/or Graphene: Synthesis and Applications. Chemical Society Reviews, 42 (2013) 794–830. https://doi.org/10.1039/c2cs35353a.

[60] S. Alwarappan, A. Erdem, C. Liu, & C.-Z. Li, Probing The Electrochemical Properties of Graphene Nanosheets for Biosensing Applications. The Journal of Physical Chemistry C, 113 (2009) 8853–8857. https://doi.org/10.1021/jp9010313.

[61] S. Alwarappan, S. Boyapalle, A. Kumar, C. Z. Li, & S. Mohapatra, Comparative Study of Single-, Few-, and Multilayered Graphene toward Enzyme Conjugation and Electrochemical Response. Journal of Physical Chemistry C, 116 (2012) 6556–6559. https://doi.org/10.1021/jp211201b.

[62] A. Karimi, A. Othman, A. Uzunoglu, L. Stanciu, & S. Andreescu, Graphene Based Enzymatic Bioelectrodes and Biofuel Cells. Nanoscale, 7 (2015) 6909–6923. https://doi.org/10.1039/c4nr07586b.

[63] L. Yan, Y. B. Zheng, F. Zhao, S. Li, X. Gao, B. Xu, P. S. Weiss, & Y. Zhao, Chemistry and Physics of A Single Atomic Layer: Strategies and Challenges for Functionalization of Graphene and Graphene-Based Materials. Chemical Society Reviews, 41 (2012) 97–114. https://doi.org/10.1039/c1cs15193b.

[64] K. P. Loh, Q. Bao, P. K. Ang, & J. Yang, The Chemistry of Graphene. Journal of
      Materials Chemistry, 20 (2010) 2277. https://doi.org/10.1039/b920539j.

[65] J. P. Smith, C. W. Foster, J. P. Metters, O. B. Sutcliffe, & C. E. Banks, Metallic
      Impurities in Graphene Screen-Printed Electrodes Can Influence Their
      Electrochemical Properties. Electroanalysis, 26 (2014) 2429–2433.
      https://doi.org/10.1002/elan.201400320.

[66] F. Banhart, J. Kotakoski, & A. V. Krasheninnikov, Structural Defects in Graphene
      RID A-3473-2009. Acs Nano, 5 (2011) 26–41.
      https://doi.org/10.1021/nn102598m.

[67] R. Singh & P. Kroll, Magnetism in Graphene due to Single-Atom Defects:
      Dependence on The Concentration and Packing Geometry of Defects. Journal of
      Physics: Condensed Matter, 21 (2009) 196002–9. https://doi.org/10.1088/0953-
      8984/21/19/196002.

[68] D. W. Boukhvalov & M. I. Katsnelson, Chemical Functionalization of Graphene.
      Journal of Physics: Condensed Matter, 21 (2009) 344205.
      https://doi.org/10.1088/0953-8984/21/34/344205.

[69] W. Wei & X. Qu, Extraordinary Physical Properties of Functionalized Graphene.
      Small, 8 (2012) 2138–2151. https://doi.org/10.1002/smll.201200104.

[70] V. Georgakilas, A. B. Bourlinos, R. Zboril, T. A. Steriotis, P. Dallas, A. K. Stubos,
      & C. Trapalis, Organic Functionalisation of Graphenes. Chemical
      Communications, 46 (2010) 1766–1768. https://doi.org/10.1039/b922081j.

[71] M. Quintana, K. Spyrou, M. Grzelczak, W. R. Browne, P. Rudolf, & M. Prato,
      Functionalization of Graphene via 1,3-Dipolar Cycloaddition. ACS Nano, 4 (2010)
      3527–3533.

[72] H. Bai, Y. Xu, L. Zhao, C. Li, & G. Shi, Non-covalent Functionalization of
      Graphene Sheets by Sulfonated Polyaniline. Chemical Communications, 0 (2009)
      1667–1669. https://doi.org/10.1039/b821805f.

[73] C. Xu, X. Wang, & J. Zhu, Graphene−Metal Particle Nanocomposites. The
      Journal of Physical Chemistry C, 112 (2008) 19841–19845.
      https://doi.org/10.1021/jp807989b.

[74] V. Georgakilas, M. Otyepka, A. B. Bourlinos, V. Chandra, N. Kim, K. C. Kemp,
      P. Hobza, R. Zboril, & K. S. Kim, Functionalization of Graphene: Covalent and
      Non-Covalent Approach. Chemical Reviews, 112 (2012) 6156–6214.
      https://doi.org/10.1021/cr3000412.

[75] J. Shen, M. Shi, B. Yan, H. Ma, N. Li, Y. Hu, & M. Ye, Covalent Attaching
      Protein to Graphene Oxide via Diimide-Activated Amidation. Colloids and

Surfaces B: Biointerfaces, 81 (2010) 434–438.
https://doi.org/10.1016/j.colsurfb.2010.07.035.

[76]   E. Demir, B. Sen, & F. Sen, Highly Efficient Pt Nanoparticles and f-MWCNT
        Nanocomposites Based Counter Electrodes for Dye-Sensitized Solar Cells. Nano-
        Structures & Nano-Objects, 11 (2017) 39–45.
        https://doi.org/10.1016/j.nanoso.2017.06.003.

[77]   Y. Yıldız, E. Erken, H. Pamuk, H. Sert, & F. Şen, Monodisperse Pt Nanoparticles
        Assembled on Reduced Graphene Oxide: Highly Efficient and Reusable Catalyst
        for Methanol Oxidation and Dehydrocoupling of Dimethylamine-Borane
        (DMAB). Journal of Nanoscience and Nanotechnology, 16 (2016) 5951–5958.
        https://doi.org/10.1166/jnn.2016.11710.

[78]   S. Calabrese Barton, J. Gallaway, & P. Atanassov, Enzymatic Biofuel Cells for
        Implantable and Microscale Devices. Chemical Reviews, 104 (2004) 4867–4886.
        https://doi.org/10.1021/cr020719k.

[79]   S. D. Minteer, B. Y. Liaw, & M. J. Cooney, Enzyme-Based Biofuel Cells. Current
        Opinion in Biotechnology, 18 (2007) 228–234.
        https://doi.org/10.1016/j.copbio.2007.03.007.

[80]   V. Flexer, N. Brun, R. Backov, & N. Mano, Designing Highly Efficient Enzyme-
        Based Carbonaceous Foams Electrodes For Biofuel Cells. Energy and
        Environmental Science, 3 (2010) 1302–1306. https://doi.org/10.1039/c003488f.

[81]   M. H. Osman, A. A. Shah, & F. C. Walsh, Recent Progress and Continuing
        Challenges in Bio-Fuel Cells. Part I: Enzymatic Cells. Biosensors and
        Bioelectronics, 26 (2011) 3087–3102. https://doi.org/10.1016/j.bios.2011.01.004.

[82]   W. Zheng, H. Y. Zhao, J. X. Zhang, H. M. Zhou, X. X. Xu, Y. F. Zheng, Y. B.
        Wang, Y. Cheng, & B. Z. Jang, A Glucose/O2 Biofuel Cell Base on
        Nanographene Platelet-Modified Electrodes. Electrochemistry Communications,
        12 (2010) 869–871. https://doi.org/10.1016/j.elecom.2010.04.006.

[83]   M. Zayats, B. Willner, & I. Willner, Design of Amperometric Biosensors and
        Biofuel Cells by The Reconstitution of Electrically Contacted Enzyme Electrodes.
        Electroanalysis, 20 (2008) 583–601. https://doi.org/10.1002/elan.200704128.

[84]   J. N. Talbert & J. M. Goddard, Enzymes on material surfaces. Colloids and
        Surfaces B: Biointerfaces, 93 (2012) 8–19.
        https://doi.org/10.1016/j.colsurfb.2012.01.003.

[85]   C. Mateo, J. M. Palomo, G. Fernandez-Lorente, J. M. Guisan, & R. Fernandez-
        Lafuente, Improvement of Enzyme Activity, Stability and Selectivity via

Immobilization Techniques. Enzyme and Microbial Technology, 40 (2007) 1451–1463. https://doi.org/10.1016/j.enzmictec.2007.01.018.

[86] D. A. J. Rand & R. M. Dell, The Hydrogen Economy: A Threat or An Opportunity for lead–Acid Batteries? Journal of Power Sources, 144 (2005) 568–578. https://doi.org/10.1016/j.jpowsour.2004.11.017.

[87] J. Kim, J. W. Grate, & P. Wang, Nanobiocatalysis and Its Potential Applications. Trends in Biotechnology, 26 (2008) 639–646. https://doi.org/10.1016/j.tibtech.2008.07.009.

[88] S. A. Ansari & Q. Husain, Potential Applications of Enzymes Immobilized on/in Nano Materials: A Review. Biotechnology Advances, 30 (2012) 512–523. https://doi.org/10.1016/j.biotechadv.2011.09.005.

[89] C. Liu, S. Alwarappan, Z. Chen, X. Kong, & C. Z. Li, Membraneless Enzymatic Biofuel Cells Based on Graphene Nanosheets. Biosensors and Bioelectronics, 25 (2010) 1829–1833. https://doi.org/10.1016/j.bios.2009.12.012.

[90] C. Liu, Z. Chen, & C. Z. Li, Surface Engineering of Graphene-Enzyme Nanocomposites for Miniaturized Biofuel Cell. IEEE Transactions on Nanotechnology, 10 (2011) 59–62. https://doi.org/10.1109/TNANO.2010.2050147.

[91] İ. Gulçin, P. Taslimi, A. Aygün, N. Sadeghian, E. Bastem, O. I. Kufrevioglu, F. Turkan, & F. Şen, Antidiabetic and Antiparasitic Potentials: Inhibition Effects of Some Natural Antioxidant Compounds on α-glycosidase, α-amylase and Human Glutathione S-transferase Enzymes. International Journal of Biological Macromolecules, 119 (2018) 741–746. https://doi.org/10.1016/j.ijbiomac.2018.08.001.

[92] B. Aday, H. Pamuk, M. Kaya, & F. Sen, Graphene Oxide as Highly Effective and Readily Recyclable Catalyst Using for The One-Pot Synthesis of 1,8-Dioxoacridine Derivatives. Journal of Nanoscience and Nanotechnology, 16 (2016) 6498–6504. https://doi.org/10.1166/jnn.2016.12432.

[93] B. Sen, E. Kuyuldar, B. Demirkan, T. Onal Okyay, A. Şavk, & F. Sen, Highly Efficient Polymer Supported Monodisperse Ruthenium-Nickel Nanocomposites For Dehydrocoupling of Dimethylamine Borane. Journal of Colloid and Interface Science, 526 (2018) 480–486. https://doi.org/10.1016/j.jcis.2018.05.021.

[94] B. Sen, S. Kuzu, E. Demir, S. Akocak, & F. Sen, Polymer-Graphene Hybrid Decorated Pt Nanoparticles as Highly Efficient And Reusable Catalyst for The Dehydrogenation of Dimethylamine–Borane at Room Temperature. International

Journal of Hydrogen Energy, 42 (2017) 23284–23291.
https://doi.org/10.1016/j.ijhydene.2017.05.112.

[95]   Y. Yıldız, İ. Esirden, E. Erken, E. Demir, M. Kaya, & F. Şen, Microwave (Mw)-
       assisted Synthesis of 5-Substituted 1H-Tetrazoles via [3+2] Cycloaddition
       Catalyzed by Mw-Pd/Co Nanoparticles Decorated on Multi-Walled Carbon
       Nanotubes. ChemistrySelect, 1 (2016) 1695–1701.
       https://doi.org/10.1002/slct.201600265.

[96]   H. Uk Lee, H. Young Yoo, T. Lkhagvasuren, Y. Seok Song, C. Park, J. Kim, & S.
       Wook Kim, Enzymatic Fuel Cells Based on Electrodeposited Graphite
       Oxide/Cobalt Hydroxide/Chitosan Composite-Enzyme Electrode. Biosensors and
       Bioelectronics, 42 (2013) 342–348. https://doi.org/10.1016/j.bios.2012.10.020.

[97]   M. W. Pitcher, Y. Arslan, P. Edinç, M. Kartal, M. Masjedi, Ö. Metin, F. Şen, Ö.
       Türkarslan, & B. Yiğitsoy, Recent Advances in The Synthesis and Applications of
       Inorganic Polymer. Phosphorus, Sulfur and Silicon and the Related Elements, 182
       (2007) 2861–2880. https://doi.org/10.1080/10426500701540431.

[98]   J. Zhang, M. P. Landry, P. W. Barone, J. H. Kim, S. Lin, Z. W. Ulissi, D. Lin, B.
       Mu, A. A. Boghossian, A. J. Hilmer, A. Rwei, A. C. Hinckley, S. Kruss, M. A.
       Shandell, N. Nair, S. Blake, F. Şen, S. Şen, R. G. Croy, D. Li, K. Yum, J. H. Ahn,
       H. Jin, D. A. Heller, J. M. Essigmann, D. Blankschtein, & M. S. Strano, Molecular
       Recognition Using Corona Phase Complexes Made of Synthetic Polymers
       Adsorbed on Carbon Nanotubes. Nature Nanotechnology, 8 (2013) 959–968.
       https://doi.org/10.1038/nnano.2013.236.

[99]   B. Devadas, V. Mani, & S. M. Chen, A Glucose/O2 Biofuel Cell Based on
       Graphene and Multiwalled Carbon Nanotube Composite Modified Electrode.
       International Journal of Electrochemical Science, 7 (2012) 8064–8075.

[100] Ö. Karatepe, Y. Yıldız, H. Pamuk, S. Eris, Z. Dasdelen, F. Sen, Enhanced
       Electrocatalytic Activity and Durability of Highly Monodisperse Pt@PPy–PANI
       Nanocomposites as A Novel Catalyst for The Electro-Oxidation of Methanol. RSC
       Advances, 6 (2016) 50851–50857. https://doi.org/10.1039/C6RA06210E.

[101] E. Erken, Y. Yıldız, B. Kilbaş, F. Şen, Synthesis and Characterization of Nearly
       Monodisperse Pt Nanoparticles for C_1 to C_3 Alcohol Oxidation and
       Dehydrogenation of Dimethylamine-borane (DMAB). Journal of Nanoscience and
       Nanotechnology, 16 (2016) 5944–5950. https://doi.org/10.1166/jnn.2016.11683.

[102] E. Erken, I. Esirden, M. Kaya, F. Sen, A Rapid and Novel Method for The
       Synthesis Of 5-Substituted 1H-Tetrazole Catalyzed by Exceptional Reusable

Materials Research Forum LLC
doi: http://dx.doi.org/10.21741/9781644900079-6

Monodisperse Pt Nps@AC Under The Microwave Irradiation. RSC Advances, 5 (2015) 68558–68564. https://doi.org/10.1039/c5ra11426h.

[103] F. Şen, G. Gökağaç, Pt nanoparticles Synthesized with New Surfactants: Improvement In C1-C3alcohol Oxidation Catalytic Activity. Journal of Applied Electrochemistry, 44 (2014) 199–207. https://doi.org/10.1007/s10800-013-0631-5.

[104] F. Şen, G. Gökağaç, S. Şen, High Performance Pt Nanoparticles Prepared by New Surfactants for C 1 To C3 Alcohol Oxidation Reactions. Journal of Nanoparticle Research, 15 (2013) 1979. https://doi.org/10.1007/s11051-013-1979-5.

[105] I. V. Pavlidis, M. Patila, U. T. Bornscheuer, D. Gournis, & H. Stamatis, Graphene-Based Nanobiocatalytic Systems: Recent Advances and Future Prospects. Trends in Biotechnology, 32 (2014) 312–320. https://doi.org/10.1016/j.tibtech.2014.04.004.

[106] J. Filip & J. Tkac, Is Graphene Worth Using in Biofuel Cells? Electrochimica Acta, 136 (2014) 340–354. https://doi.org/10.1016/j.electacta.2014.05.119.

[107] B. Unnikrishnan, S. Palanisamy, & S. M. Chen, A Simple Electrochemical Approach to Fabricate A Glucose Biosensor Based on Graphene-Glucose Oxidase Biocomposite. Biosensors and Bioelectronics, 39 (2013) 70–75. https://doi.org/10.1016/j.bios.2012.06.045.

[108] S. Palanisamy, S. Cheemalapati, & S. M. Chen, An enzymatic Biofuel Cell Based on Electrochemically Reduced Graphene Oxide and Multiwalled Carbon Nanotubes/Zinc Oxide Modified Electrode. International Journal of Electrochemical Science, 7 (2012) 11477–11487.

[109] Z. Chen, W. Ren, L. Gao, B. Liu, S. Pei, & H. M. Cheng, Three-Dimensional Flexible and Conductive Interconnected Graphene Networks Grown by Chemical Vapour Deposition. Nature Materials, 10 (2011) 424–428. https://doi.org/10.1038/nmat3001.

[110] Y. Zhang, M. Chu, L. Yang, Y. Tan, W. Deng, M. Ma, X. Su, & Q. Xie, Three-Dimensional Graphene Networks as A New Substrate for Immobilization of Laccase and Dopamine And Its Application in Glucose/O2 Biofuel Cell. ACS Applied Materials and Interfaces, 6 (2014) 12808–12814. https://doi.org/10.1021/am502791h.

[111] K. P. Prasad, Y. Chen, & P. Chen, Three-Dimensional Graphene-Carbon Nanotube Hybrid for High-Performance Enzymatic Biofuel Cells. ACS Applied Materials and Interfaces, 6 (2014) 3387–3393. https://doi.org/10.1021/am405432b.

[112] Y. Koskun, A. Şavk, B. Şen, & F. Şen, Highly Sensitive Glucose Sensor Based on Monodisperse Palladium Nickel/Activated Carbon Nanocomposites. Analytica Chimica Acta, 1010 (2018) 37–43. https://doi.org/10.1016/j.aca.2018.01.035.

[113] G. Başkaya, Y. Yıldız, A. Savk, T. O. Okyay, S. Eriş, H. Sert, & F. Şen, Rapid, Sensitive, and Reusable Detection of Glucose by Highly Monodisperse Nickel Nanoparticles Decorated Functionalized Multi-Walled Carbon Nanotubes. Biosensors and Bioelectronics, 91 (2017) 728–733. https://doi.org/10.1016/j.bios.2017.01.045.

[114] G. Gökağaç, M. Sonsuz, F. Şen, & D. Kisakürek, Atom Transfer Rearrangement Radical Polymerization of Diammine-bis(2,4,6- trihalophenolato)Copper(II) Complexes in The Solid State. Zeitschrift fur Naturforschung - Section B Journal of Chemical Sciences, 61 (2006) 1222–1228.

[115] Y. Yildiz, T. O. Okyay, B. Sen, B. Gezer, S. Kuzu, A. Savk, E. Demir, Z. Dasdelen, H. Sert, & F. Sen, Highly Monodisperse Pt/Rh Nanoparticles Confined in the Graphene Oxide for Highly Efficient and Reusable Sorbents for Methylene Blue Removal from Aqueous Solutions. ChemistrySelect, 2 (2017) 697–701. https://doi.org/10.1002/slct.201601608.

Enzymatic Fuel Cells
Materials Research Foundations **44** (2019) 157-172

Materials Research Forum LLC
doi: http://dx.doi.org/10.21741/9781644900079-7

Chapter 7

Mesoporous Materials in Biofuel Cells

Mehmet Harbi Calimli[1,2], Mehmet Salih Nas[1,3], Hakan Burhan[1], Fatih Sen[1,*]

[1]Sen Research Group, Department of Biochemistry, Dumlupınar University, 43100 Kütahya, Turkey

[2]Tuzluca vocational school, Igdir University, Igdir, Turkey

[3]Engineering Faculty, Environmental Engineering Department, Igdir University, Igdir, Turkey

fatihsen1980@gmail.com

Abstract

In biological systems, the conversion of chemical energy into electrical energy exhibited the importance of biological fuel cells. Thus, studies on the use of microbes and enzymes in fuel cell devices have recently increased. Reduced fossil resources and damage to the environment have led to an increase in the search for new energy sources. Electric energy can be obtained by using microbial and enzymes in fuel cells. The efficiency and durability of the material to be used in cathodic and anodic electrodes enable the efficient use of the biofuel cell. It is very important to know the properties of the membrane materials that provide ion transfer between the cathode and the anode. So, the characteristics of the materials used in these fuel cells, which can be expressed as biological fuel cells, affect the energy efficiency. This chapter is aimed to report recent developments in microbial fuel cells, enzymatic fuel cells and carbon-based fuel cells and the materials used in these fuel cells.

Keywords

Biofuel Cells, Enzymatic Fuel Cells, Mesoporous Materials, Microbial Fuel Cells, Nanomaterials

Contents

1. Introduction

The demand for modern technologies with high performance is increasing day by day due to the limited availability of fossil resources. The innovative technologies are high performance and environmentally friendly. So, there is considerable interest in these technologies to be implemented [1–2]. Examples of the main energy sources used in modern technologies are biohydrogen [3], bioethanol [4], biodiesel [5] and enzymatic fuel cells (EFCs) [6–7]. Among these energy sources, microbial fuel cells (MFCs) and enzymatic fuel cells (EFCs) have been implemented successfully in the energy and clean environment. Microbial fuel cells or enzymatic fuel cells (EFCs) are composed of (italic) biohydrogen, biosensor and *an in-situ* power source for wastewater cleaning and biodegradation [8]. Microbial fuel cells and biological fuel cells have a technology that can convert the energy of the chemical bonds into electricity by the catalytic reactions created by microorganisms [9]. To generate electricity from inorganic and organic compounds, bacteria are used as catalysts in microbial fuel cells. During electricity generation, the compounds are oxidized by the bacteria in a microbial fuel cell. Microbes as substrates in the anionic solution of the microbial fuel cells release protons and electrons. Carbon dioxide is formed by oxidation, but carbon dioxide is not generated because it is used in the photosynthesis. Hence, carbon dioxide emissions do not occur. The decomposition of the organic material into protons and electrons is provided by the bacteria on the anode. The formed electrons flow towards the anode and then flow towards the cathode through a conductive material attached to the anode. The resulting electric current is used in an electrically operated device [9–10].

Materials Research Forum LLC

doi: http://dx.doi.org/10.21741/9781644900079-7

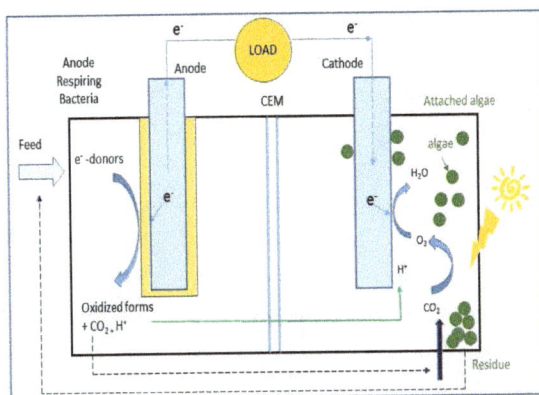

Figure 1. Schematic representation of the biological fuel cells and the formation of CO_2 and H_2O.

Microbial fuel cells and enzymatic fuel cells termed as biological fuel cells (BFCs) are found in several types, and these are generally composed of three sections, which are an anode electrode, a cathode electrode and a membrane which is between the electrodes. The membrane between the electrodes provides a proton transition. The anode has a high H^+ concentration in the section where the electrode is located. Hence H^+ ions flow through the membrane to the cathode [11].

Some algae and other microbial organisms can produce oxygen and assimilate CO_2, which makes them advantageous for use in microbial fuel cells. Bacteria and algae in the microbial fuel cells oxidize the substrates in the anode chamber, and oxygen is formed by the bacteria in the cathode chamber. The resulting oxygen absorbs the electrons in the MFC medium and reduces the amount of generated CO_2. Thus, the photosynthesis of the bacteria increases the amount of biomass and water. The chemical reactions of the microbial fuel cells are given below

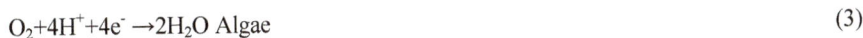

$$CO_2 + 12H^+ + 12e^- \rightarrow C_6H_{12}O_6 \text{ (biomass)} + 3O_2 \tag{1}$$

$$C_6H_{12}O_6 \text{ (biomass)} + 6O_2 \rightarrow 6CO_2 + 6H_2O \tag{2}$$

$$O_2 + 4H^+ + 4e^- \rightarrow 2H_2O \text{ Algae} \tag{3}$$

Enzymatic Fuel Cells   Materials Research Forum LLC
Materials Research Foundations **44** (2019) 157-172   doi: http://dx.doi.org/10.21741/9781644900079-7

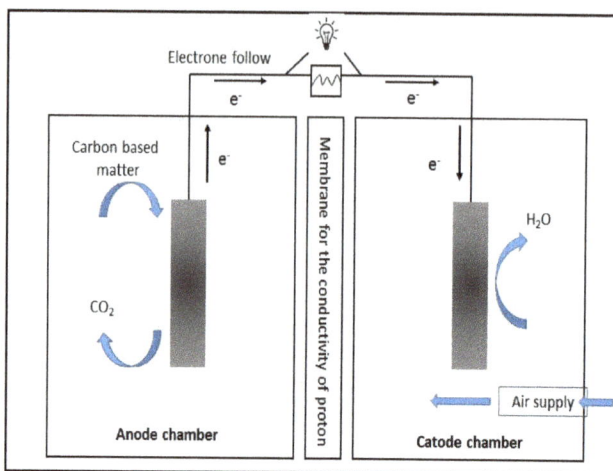

Figure 2. The structure for cathode and anode electrodes in fuel cells.

2. Electrode types and electrode construction materials in fuel cell

The main noble metals used in electrochemical applications are Au (gold), Pt (platinum), Ag (silver) and Pd (silver). There is a tendency to use low-cost metals such as Fe (iron), Cu (copper), and Ag (silver). In contrast, low-cost metals undergo corrosion in aqueous media. To overcome this negative situation, carbon-based electrodes are being tried to be developed. Common carbon-based materials used in anode electrodes are carbon cloth, carbon paper, graphite granules, and graphite. These materials are preferred due to their advantages such as chemical stability, large surface areas and low cost [1–12]. To minimize this disadvantage of noble metals, researchers use carbon nanotubes. Generally, carbon nanotubes are used in power generation, and graphene is used as an electrode [13]. To fabricate graphene oxide electrodes, several methods such as chemical derived methods, chemical vapour deposition, cleavage, and exfoliation have been used. The main purpose in all of these techniques is to prepare the chemical derivatives of graphite and to use the composite materials formed in new fields [14]. The major ones being bio-reduction, chemical addition, metal functionalization, electrochemical reduction and precipitation, etc. In all these methods, graphene oxide is obtained from graphene. In the Hummer method, graphene oxide is obtained in the form of flakes from graphite in three stages [15]. Hydrazine or sodium borane is used as the reducing agent, and at the end, the prepared graphene oxide is covered on the electrodes. Application of planar thin layer

Materials Research Forum LLC
doi: http://dx.doi.org/10.21741/9781644900079-7

graphene onto metal surface was performed using chemical vapour deposition method (CVD) by Singh. In chemical vapour deposition, applications enhanced plasma CVD, thermal decomposition, and thermal CVD are performed. This method is very common nowadays. The CVD method is considered to be widely used in the future for the manufacturing of single and thin layer graphene. In this method, the graphene is subjected to chemical treatment, isolated from environmental influences, finally removed and transferred to another substrate. At the end of the CVD method resulting graphene is obtained larger quantity than the other methods [16–18].

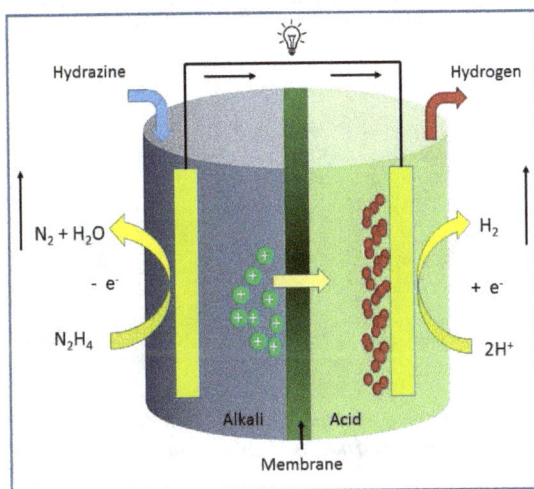

Figure 3: Schematic representation of the fuel cell in which hydrazine is used.

3. Components of Enzymatic Fuel Fells (EFCs)

Fuel cells are composed of non-selective metals or metal alloys that oxidize the substrate and reduce the oxidizing agent. Recent fuel cell investigations have focused on the development of electrolytes that can inhibit surface poisoning, require a low amount of catalyst and prevent fuel and oxidant passages. Enzymatic fuel cells (EFC) provide oxidation at the anode of the fuel chamber, while in the cathode chamber it causes reduction of the oxidant. With the use of enzymatic fuels, chemical fuel cell reactions take place very easily under room conditions. Electric energy is generated by the enzymatic degradation of glucose and oxygen in the enzymatic fuel cells. Because of the form of glucose at the end of human body metabolic events, in theory, glucose is

Enzymatic Fuel Cells   Materials Research Forum LLC
Materials Research Foundations **44** (2019) 157-172   doi: http://dx.doi.org/10.21741/9781644900079-7

regarded as an unlimited energy source [19]. The power generated in an enzymatic fuel cell depends on the current and voltage of the cell. The main factors affecting cell voltage are oxidant and fuel, cell resistance, electron flow and current state. The cell voltage of the complete oxidation of cellulose to water and carbon dioxide can be calculated to be 1.24 V using Gibbs energy at room temperature. This value is the highest voltage expecting from EFC and without any energy loss under ambient conditions [20].

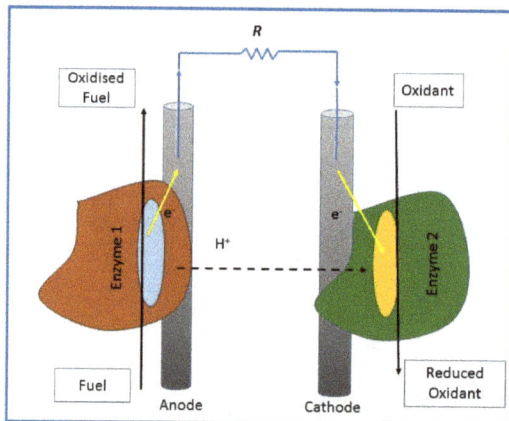

Figure 4: The schematic of the enzymatic fuel cell consists of the cathode electrode, the anode electrode, and enzyme structures.

In redox reactions occurring in electro-enzymes systems containing two components called the cofactor (small nonprotein structure) and the apoenzymes (protein structure). Electron transfer between the substrate and the enzyme is provided by the cofactor. The cofactor may be bound by strong bonds by the enzyme during the reaction or break. The major cofactors commonly used in enzyme oxidation of glucose are nicotinamide adenine dinucleotide (NAD) or flavin adenine dinucleotide (FAD). If there is a tight bond between the cofactors and the enzyme at the active sites, the transfer of the electrons from the enzyme to the electrode can take place directly. The physical environment of EFCs (in the case of glucose systems) designed as semiconductor systems for the energy field must be in appropriate conditions. These conditions are pH 7.4, temperature 37 °C, glucose and oxygen concentration 10^{-4} M and bulk transfer 1-10 cm/s. Due to the extreme sensitivity of the enzymes and films, not much work had been done under the above conditions. The few studies carried out these circumstances have been provided by

Enzymatic Fuel Cells   Materials Research Forum LLC
Materials Research Foundations **44** (2019) 157-172   doi: http://dx.doi.org/10.21741/9781644900079-7

immobilization of enzymes. Personal power production has been tried to be achieved by using body fluids such as saliva, tears, and sweat [21].

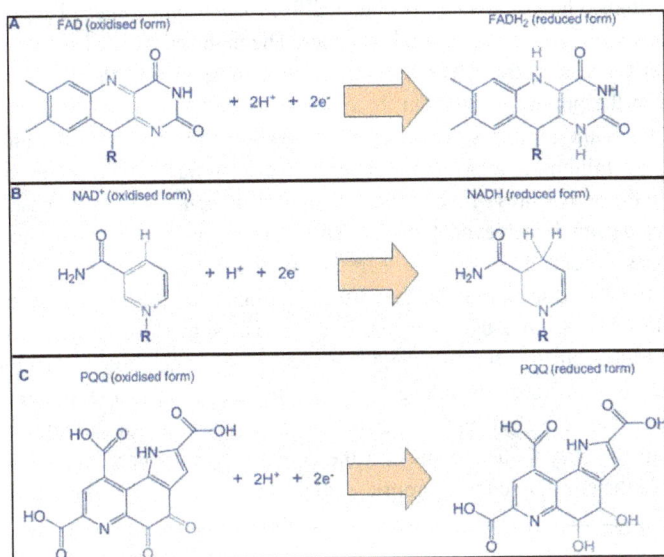

Figure 5: The cofactors found in enzymatic fuel cells and the resulting redox reaction.

3.1   Cathode electrode and its process in enzymatic and microbial fuel cells

The cathodes in the enzymatic and microbial fuel cells consisting of a multicopper enzyme family four copper centers. The mononuclear copper in the cathode provides electron transfer between the copper center electrode and the substrate to reduce the oxygen into water. The multi-copper-based electrodes have been developed based on the fact that enzymes are applied to the electrodes in a controlled manner. A center located on the multi-copper-based cathode binds phenolic functional groups, activates oxygen and allows reduction to occur. Copper tips that provide oxidation are positioned close to the protein constructs and thus provide an orientation of the enzymes. As oxidases bilirubin and laccases are commonly used due to their high redox potential for the copper center in cathode electrode [22]. As described above, graphite-based electrodes are also widely used because of their high performance in the direct electrical wiring of laccases and immobilizations. Most of the electrodes used today are based on carbon nanotubes [23].

Materials Research Forum LLC

doi: http://dx.doi.org/10.21741/9781644900079-7

3.2 Anode electrode and their process in enzymatic and microbial fuel cells

Glucose oxidase has high activity, stability, and specificity against b - D-glucose in biological fluids. Glucose oxidase is obtained from *Aspergillus niger* (Tiegh., 1867) and is a glucose oxidizing biocatalyst for bioanodes. The high activity and stability of glucose oxidase exist because the active site in the protein shell and the FAD cofactor are embedded in the protein matrix. On the other hand, embedding the active side into the deep regions of the protein structure by the FAD cofactor makes it difficult to transfer electrons. To minimize these disadvantages, electron transfer is achieved by redox reactions in the active sites of the enzyme using carbon nanotube (CNT)-based materials, and biofuel design is achieved [24–25]. The relationship between enzyme and anode electrode was simulated in Fig. 6. If there is a tight bond between the cofactors and the enzyme at the active sites, the transfer of the electrons from the enzyme to the electrode can take place directly. Though many enzymes have been tried for direct electron transfer between the electrode and the enzyme, there exist many difficulties to obtain an efficient current between the enzyme active site and the electrode [26–27]. Some of these difficulties include the electron transfer that can take place only at certain distances [21–28] between the enzyme active site and the cofactor and the necessity of directing the active site of the enzyme to the appropriate side.

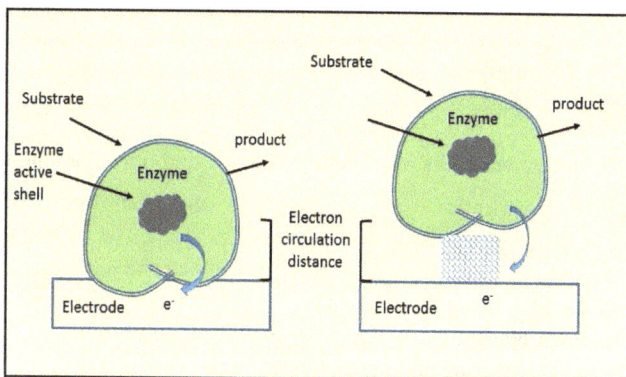

Figure 6. The relationship between the enzyme and the anode electrode.

3.3 Membrane and membrane properties in fuel cell

The membrane is located between the anode and the cathode in the fuel cells. The effectiveness of the membranes used in the fuel cells depends on several factors such as

Materials Research Forum LLC
doi: http://dx.doi.org/10.21741/9781644900079-7

the type of material, working medium, membrane modification, and membrane preparation method. Nowadays polyamides and their derivations as thin film composites (TFC) in membranes have wide use due to high water permeability, salt rejection, wide operating temperature range (1-45 °C) and appropriate pH tolerance (1-11) [29]. A typical TFC consists of three parts; these are a thick layer of polyamide, a substrate of polysulfone or polyether polysulfone, and a layer of fabric providing mechanical resistance [30]. The structure of TFC is simulated as shown in Figure 7.

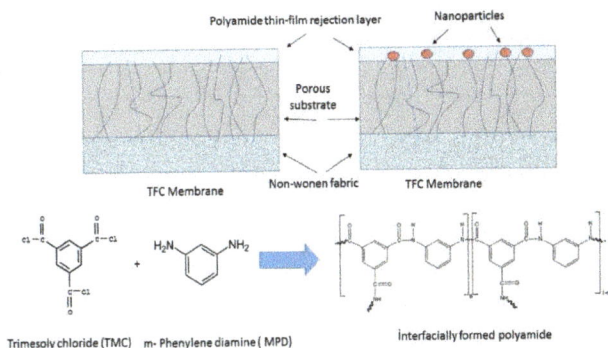

Figure 7. The structure of the TFC membrane consisted of three sections.

Another type of membrane is polymer electrolyte and graphene sheets. Polymer graphene sheets membranes are having some properties such as membrane durability, water permeability, hydraulic-thermal resistance, fuel transport resistance, high mechanical strength and electrochemical-chemical stability, and these make them preferrable in a fuel cell. Membranes with the desired properties can be synthesized by functionalization of the graphene. Thus, recent studies have concentrated on the functionalization of graphite. Graphene oxide (GO) and reduced graphene (r-GO) oxide can be obtained by functionalization of graphite using the Hummer's method. The resulting structure of graphite functionalization has been simulated in the following image. Graphene oxide is a structure composed of functional groups such as phenyl, epoxy, carboxyl and hydroxyl [31].

Materials Research Forum LLC
doi: http://dx.doi.org/10.21741/9781644900079-7

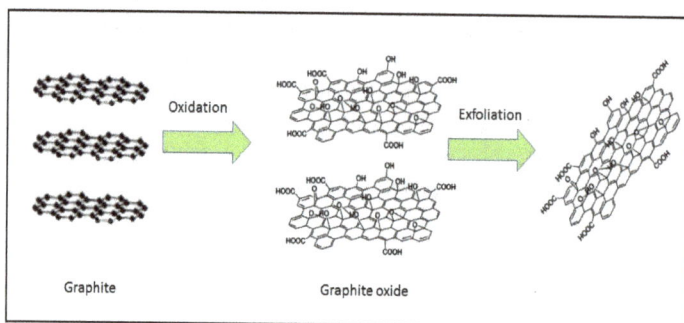

Figure 8. The forming GO from graphite and gaining some functional groups.

3.4   Carbon Based Materials (CBMs)

Carbon-based materials (CBMs) have been used as conductive substrates in enzymatic biosystems due to their large high potential. Their high surface areas, unique mechanical-electronic properties that facilitate direct electron transfer enhancement and mediator electron transport in enzymes are among the reasons for their preference [32]. The CBMs are well suited to improve stability and electronic properties by bonding two functional group structures. Also, CBMs have been modified using appropriate functional groups. Bis-pyrene-2,20-azino-bis compound is a functional group used for this purpose [25–33]. By attaching carbon-based materials to such functional groups, CBMs can be significantly improved in properties such as stability, matrix alignment, and electrocatalytic properties [21–29–31–34–37]. Carbon-based materials are among the leading materials for forming electrocatalytic bioelectrodes [38–45]. The membrane used in a fuel cell having some properties such as low cost, high durability, and high performance are preferred. The main function of the membranes is to be between the cathode and the anode, to ensure ion transport and to provide fuel insulation [46–54].

Conclusion

Microbial fuel cells and enzymatic fuel cells can be used efficiently in the clean energy field. Factors such as the durability, conductivity, and efficiency of the materials used in biological fuel cells affect the usability of the battery. Some disadvantages of fuel cells can be eliminated by using inexpensive carbon-based materials instead of high-cost noble metals used in cathodes and anodes. With high surface area, easy electron transport

Materials Research Forum LLC
doi: http://dx.doi.org/10.21741/9781644900079-7

capacity and other superior features, carbon nanotubes can be easily used in biological fuel cells. In this chapter, general information about biological fuel cells and the best knowledge about the electrodes and membranes used in these cells have been examined. Additionally, recent developments in carbon-based carbon nanotubes have been examined. Recently, the prohibition of diesel-fueled automobiles in some countries has increased the importance of studies on electric-powered automobiles. Efficiency in fuel cells for efficient and clean energy in the future depends on the efficiency of the materials used.

References

[1]. A. ElMekawy, H. M. Hegab, D. Losic, C. P. Saint, & D. Pant, Applications of Graphene in Microbial Fuel Cells: The Gap between Promise and Reality. Renewable and Sustainable Energy Reviews, 72 (2017) 1389–1403. https://doi.org/10.1016/j.rser.2016.10.044.

[2]. K. Zhang, L. L. Zhang, X. S. Zhao, & J. Wu, Graphene/Polyaniline Nanofiber Composites as Supercapacitor Electrodes. Chemistry of Materials, 22 (2010) 1392–1401. https://doi.org/10.1021/cm902876u.

[3]. M. Chan Sin, S. N. Gan, M. S. Mohd Annuar, & I. K. Ping Tan, Thermodegradation of Medium-Chain-Length Poly(3-Hydroxyalkanoates) Produced by Pseudomonas Putida from Oleic Acid. Polymer Degradation and Stability, 95 (2010) 2334–2342. https://doi.org/10.1016/j.polymdegradstab.2010.08.027.

[4]. A. Parmar, N. K. Singh, A. Pandey, E. Gnansounou, & D. Madamwar, Cyanobacteria and Microalgae: A Positive Prospect for Biofuels. Bioresource Technology, 102 (2011) 10163–10172. https://doi.org/10.1016/j.biortech.2011.08.030.

[5]. E. M. Ahmed, Hydrogel: Preparation, Characterization, and Applications: A Review. Journal of Advanced Research, 6 (2015) 105–121. https://doi.org/10.1016/j.jare.2013.07.006.

[6]. D. R. Lovley, Bug juice: Harvesting Electricity with Microorganisms. Nature Reviews Microbiology, 4 (2006) 497–508. https://doi.org/10.1038/nrmicro1442.

[7]. Y. Hindatu, M. S. M. Annuar, & A. M. Gumel, Mini-review: Anode Modification for Improved Performance of Microbial Fuel Cell. Renewable and Sustainable Energy Reviews, 73 (2017) 236–248. https://doi.org/10.1016/j.rser.2017.01.138.

[8]. L. He, P. Du, Y. Chen, H. Lu, X. Cheng, B. Chang, & Z. Wang, Advances in Microbial Fuel Cells for Wastewater Treatment. Renewable and Sustainable Energy Reviews, 71 (2017) 388–403. https://doi.org/10.1016/j.rser.2016.12.069.

[9].   M. H. Do, H. H. Ngo, W. S. Guo, Y. Liu, S. W. Chang, D. D. Nguyen, L. D. Nghiem, & B. J. Ni, Challenges in The Application of Microbial Fuel Cells to Wastewater Treatment and Energy Production: A Mini Review. Science of The Total Environment, 639 (2018) 910–920. https://doi.org/10.1016/j.scitotenv.2018.05.136.

[10].   G. Hernández-Flores, H. M. Poggi-Varaldo, O. Solorza-Feria, M. T. Ponce-Noyola, T. Romero-Castañón, N. Rinderknecht-Seijas, & J. Galíndez-Mayer, Characteristics of A Single Chamber Microbial Fuel Cell Equipped with A Low Cost Membrane. International Journal of Hydrogen Energy, 40 (2015) 17380–17387. https://doi.org/10.1016/j.ijhydene.2015.10.024.

[11].   J. L. Brown, Fuel Cells Treat Wastewater, Generate Electricity. Civil Engineering Magazine Archive, 82 (2012) 34–35. https://doi.org/10.1061/ciegag.0000643.

[12].   G. G. kumar, V. G. S. Sarathi, & K. S. Nahm, Recent Advances and Challenges in The Anode Architecture and Their Modifications for The Applications of Microbial Fuel Cells. Biosensors and Bioelectronics, 43 (2013) 461–475. https://doi.org/10.1016/j.bios.2012.12.048.

[13].   E. Flahaut, M. C. Durrieu, M. Remy-Zolghadri, R. Bareille, & C. Baquey, Study of The Cytotoxicity of CCVD Carbon Nanotubes. Journal of Materials Science, 41 (2006) 2411–2416. https://doi.org/10.1007/s10853-006-7069-7.

[14].   Y. Yıldız, E. Erken, H. Pamuk, H. Sert, & F. Şen, Monodisperse Pt Nanoparticles Assembled on Reduced Graphene Oxide: Highly Efficient and Reusable Catalyst for Methanol Oxidation and Dehydrocoupling of Dimethylamine-Borane (DMAB). Journal of Nanoscience and Nanotechnology, 16 (2016) 5951–5958. https://doi.org/10.1166/jnn.2016.11710.

[15].   S. Akocak, B. Şen, N. Lolak, A. Şavk, M. Koca, S. Kuzu, & F. Şen, One-Pot Three-Component Synthesis of 2-Amino-4H-Chromene Derivatives by Using Monodisperse Pd Nanomaterials Anchored Graphene Oxide as Highly Efficient and Recyclable Catalyst. Nano-Structures and Nano-Objects, 11 (2017) 25–31. https://doi.org/10.1016/j.nanoso.2017.06.002.

[16].   N. Neuberger, H. Adidharma, & M. Fan, Graphene: A Review of Applications in The Petroleum Industry. Journal of Petroleum Science and Engineering, 167 (2018) 152–159. https://doi.org/10.1016/j.petrol.2018.04.016.

[17].   V. Singh, D. Joung, L. Zhai, S. Das, S. I. Khondaker, & S. Seal, Graphene Based Materials: Past, Present and Future. Progress in Materials Science, 56 (2011) 1178–1271. https://doi.org/10.1016/j.pmatsci.2011.03.003.

Materials Research Forum LLC

doi: http://dx.doi.org/10.21741/9781644900079-7

[18]. P. R. Somani, S. P. Somani, & M. Umeno, Planer Nano-Graphenes from Camphor by CVD. Chemical Physics Letters, 430 (2006) 56–59. https://doi.org/10.1016/j.cplett.2006.06.081.

[19]. M. Rasmussen, S. Abdellaoui, & S. D. Minteer, Enzymatic Biofuel Cells: 30 Years of Critical Advancements. Biosensors and Bioelectronics, 76 (2016) 91–102. https://doi.org/10.1016/j.bios.2015.06.029.

[20]. E. Simon, C. M. Halliwell, C. S. Toh, A. E. G. Cass, & P. N. Bartlett, Immobilisation of Enzymes on Poly(Aniline)-Poly(Anion) Composite Films. Preparation of Bioanodes for Biofuel Cell Applications. Bioelectrochemistry, 55 (2002) 13–15. https://doi.org/10.1016/S1567-5394(01)00160-8.

[21]. D. Leech, P. Kavanagh, & W. Schuhmann, Enzymatic Fuel Cells: Recent Progress. Electrochimica Acta, 84 (2012) 223–234. https://doi.org/10.1016/j.electacta.2012.02.087.

[22]. N. Mano & L. Edembe, Bilirubin Oxidases in Bioelectrochemistry: Features and Recent Findings. Biosensors and Bioelectronics, 50 (2013) 478–485. https://doi.org/10.1016/j.bios.2013.07.014.

[23]. E. Nazaruk, K. Sadowska, J. F. Biernat, J. Rogalski, G. Ginalska, & R. Bilewicz, Enzymatic Electrodes Nanostructured with Functionalized Carbon Nanotubes for Biofuel Cell Applications. Analytical and Bioanalytical Chemistry, 398 (2010) 1651–1660. https://doi.org/10.1007/s00216-010-4012-1.

[24]. A. Zebda, C. Gondran, A. Le Goff, M. Holzinger, P. Cinquin, & S. Cosnier, Mediatorless High-Power Glucose Biofuel Cells Based on Compressed Carbon Nanotube-Enzyme Electrodes. Nature Communications, 2 (2011) 370–376. https://doi.org/10.1038/ncomms1365.

[25]. S. Cosnier, A. J. Gross, A. Le Goff, & M. Holzinger, Recent advances on Enzymatic Glucose/Oxygen and Hydrogen/Oxygen Biofuel Cells: Achievements and Limitations. Journal of Power Sources, 325 (2016) 252–263. https://doi.org/10.1016/j.jpowsour.2016.05.133.

[26]. R. Ludwig, W. Harreither, F. Tasca, & L. Gorton, Cellobiose Dehydrogenase: A Versatile Catalyst for Electrochemical Applications. ChemPhysChem, 11 (2010) 2674–2697. https://doi.org/10.1002/cphc.201000216.

[27]. A. Christenson, N. Dimcheva, E. E. Ferapontova, L. Gorton, T. Ruzgas, L. Stoica, S. Shleev, A. I. Yaropolov, D. Haltrich, R. N. F. Thorneley, & S. D. Aust, Direct Electron Transfer Between Ligninolytic Redox Enzymes and Electrodes. Electroanalysis, 16 (2004) 1074–1092. https://doi.org/10.1002/elan.200403004.

Materials Research Forum LLC
doi: http://dx.doi.org/10.21741/9781644900079-7

[28]. R. A. Marcus & N. Sutin, Electron Transfers in Chemistry and Biology. Biochimica et Biophysica Acta (BBA) - Reviews on Bioenergetics, 811 (1985) 265–322. https://doi.org/10.1016/0304-4173(85)90014-X.

[29]. A. G. Fane, C. Y. Tang, & R. Wang, Membrane Technology for Water: Microfiltration, Ultrafiltration, Nanofiltration, and Reverse Osmosis. Treatise on Water Science, 4 (2010) 301–335. https://doi.org/10.1016/B978-0-444-53199-5.00091-9.

[30]. Z. Yang, X.-H. Ma, & C. Y. Tang, Recent Development of Novel Membranes for Desalination. Desalination, 434 (2018) 37–59. https://doi.org/10.1016/j.desal.2017.11.046.

[31]. U. R. Farooqui, A. L. Ahmad, & N. A. Hamid, Graphene oxide: A Promising Membrane Material for Fuel Cells. Renewable and Sustainable Energy Reviews, 82 (2018) 714–733. https://doi.org/10.1016/j.rser.2017.09.081.

[32]. A. A. Babadi, S. Bagheri, & S. B. A. Hamid, Progress on Implantable Biofuel Cell: Nano-Carbon Functionalization for Enzyme Immobilization Enhancement. Biosensors and Bioelectronics, 79 (2016) 850–860. https://doi.org/10.1016/j.bios.2016.01.016.

[33]. M. J. Cliffe, J. A. Hill, C. A. Murray, F.-X. Coudert, & A. L. Goodwin, Freestanding Redox Buckypaper Electrodes from Multi-Wall Carbon Nanotubes For Bioelectrocatalytic Oxygen Reduction via Mediated Electron Transfer. Chemical Science, 12 (2015) 6397–6406. https://doi.org/10.1039/b000000x.

[34]. S. Wang, R. Downes, C. Young, D. Haldane, A. Hao, R. Liang, B. Wang, C. Zhang, & R. Maskell, Carbon Fiber/Carbon Nanotube Buckypaper Interply Hybrid Composites: Manufacturing Process and Tensile Properties. Advanced Engineering Materials, 17 (2015) 1442–1453. https://doi.org/10.1002/adem.201500034.

[35]. F. Fischer, Photoelectrode, Photovoltaic and Photosynthetic Microbial Fuel Cells. Renewable and Sustainable Energy Reviews, 90 (2018) 16–27. https://doi.org/10.1016/j.rser.2018.03.053.

[36]. C. Schulz, R. Kittl, R. Ludwig, & L. Gorton, Direct Electron Transfer from the FAD Cofactor of Cellobiose Dehydrogenase to Electrodes. ACS Catalysis, 6 (2016) 555–563. https://doi.org/10.1021/acscatal.5b01854.

[37]. Z. Daşdelen, Y. Yıldız, S. Eriş, & F. Şen, Enhanced Electrocatalytic Activity and Durability of Pt Nanoparticles Decorated on GO-PVP Hybride Material for Methanol Oxidation Reaction. Applied Catalysis B: Environmental, 219 (2017) 511–516. https://doi.org/10.1016/j.apcatb.2017.08.014.

Materials Research Forum LLC
doi: http://dx.doi.org/10.21741/9781644900079-7

[38]. H. Goksu, Y. Yıldız, B. Çelik, M. Yazici, B. Kilbas, & F. Sen, Eco-Friendly Hydrogenation of Aromatic Aldehyde Compounds by Tandem Dehydrogenation of Dimethylamine-Borane in The Presence of A Reduced Graphene Oxide Furnished Platinum Nanocatalyst. Catalysis Science & Technology, 6 (2016) 2318–2324. https://doi.org/10.1039/C5CY01462J.

[39]. H. Göksu, Y. Yıldız, B. Çelik, M. Yazıcı, B. Kılbaş, & F. Şen, Highly Efficient and Monodisperse Graphene Oxide Furnished Ru/Pd Nanoparticles for the Dehalogenation of Aryl Halides via Ammonia Borane. ChemistrySelect, 1 (2016) 953–958. https://doi.org/10.1002/slct.201600207.

[40]. B. Aday, Y. Yildiz, R. Ulus, S. Eris, F. Sen, & M. Kaya, One-Pot, Efficient and Green Synthesis of Acridinedione Derivatives Using Highly Monodisperse Platinum Nanoparticles Supported with Reduced Graphene Oxide. New Journal of Chemistry, 40 (2016) 748–754. https://doi.org/10.1039/c5nj02098k.

[41]. S. Bozkurt, B. Tosun, B. Sen, S. Akocak, A. Savk, M. F. Ebeoğlugil, & F. Sen, A Hydrogen Peroxide Sensor Based on TNM Functionalized Reduced Graphene Oxide Grafted with Highly Monodisperse Pd Nanoparticles. Analytica Chimica Acta, 989 (2017) 88–94. https://doi.org/10.1016/j.aca.2017.07.051.

[42]. B. Aday, H. Pamuk, M. Kaya, & F. Sen, Graphene Oxide as Highly Effective and Readily Recyclable Catalyst Using for the One-Pot Synthesis of 1,8-Dioxoacridine Derivatives. Journal of Nanoscience and Nanotechnology, 16 (2016) 6498–6504. https://doi.org/10.1166/jnn.2016.12432.

[43]. R. Ayranci, G. Başkaya, M. Güzel, S. Bozkurt, F. Şen, & M. Ak, Carbon Based Nanomaterials for High Performance Optoelectrochemical Systems. ChemistrySelect, 2 (2017) 1548–1555. https://doi.org/10.1002/slct.201601632.

[44]. B. Çelik, G. Başkaya, H. Sert, Ö. Karatepe, E. Erken, & F. Şen, Monodisperse Pt(0)/DPA@GO Nanoparticles as Highly Active Catalysts for Alcohol Oxidation and Dehydrogenation of DMAB. International Journal of Hydrogen Energy, 41 (2016) 5661–5669. https://doi.org/10.1016/j.ijhydene.2016.02.061.

[45]. E. Erken, I. Esirden, M. Kaya, & F. Sen, A Rapid and Novel Method for The Synthesis of 5-Substituted 1H-Tetrazole Catalyzed by Exceptional Reusable Monodisperse Pt NPs@AC under The Microwave Irradiation. RSC Advances, 5 (2015) 68558–68564. https://doi.org/10.1039/c5ra11426h.

[46]. Ö. Karatepe, Y. Yıldız, H. Pamuk, S. Eris, Z. Dasdelen, & F. Sen, Enhanced Electrocatalytic Activity and Durability of Highly Monodisperse Pt@Ppy–PANI Nanocomposites as A Novel Catalyst for The Electro-Oxidation of Methanol. RSC Advances, 6 (2016) 50851–50857. https://doi.org/10.1039/C6RA06210E.

[47]. E. Erken, H. Pamuk, Ö. Karatepe, G. Başkaya, H. Sert, O. M. Kalfa, & F. Şen, New Pt(0) Nanoparticles as Highly Active and Reusable Catalysts in the C1–C3 Alcohol Oxidation and the Room Temperature Dehydrocoupling of Dimethylamine-Borane (DMAB). Journal of Cluster Science, 27 (2016) 9–23. https://doi.org/10.1007/s10876-015-0892-8.

[48]. F. Sen, Y. Karatas, M. Gulcan, & M. Zahmakiran, Amylamine Stabilized Platinum(0) Nanoparticles: Active and Reusable Nanocatalyst in The Room Temperature Dehydrogenation of Dimethylamine-Borane. RSC Advances, 4 (2014) 1526–1531. https://doi.org/10.1039/c3ra43701a.

[49]. S. Eris, Z. Daşdelen, Y. Yıldız, & F. Sen, Nanostructured Polyaniline-rGO Decorated Platinum Catalyst with Enhanced Activity and Durability for Methanol Oxidation. International Journal of Hydrogen Energy, 43 (2018) 1337–1343. https://doi.org/10.1016/j.ijhydene.2017.11.051.

[50]. Y. Yıldız, S. Kuzu, B. Sen, A. Savk, S. Akocak, & F. Şen, Different Ligand Based Monodispersed Pt Nanoparticles Decorated with rGO as Highly Active and Reusable Catalysts for The Methanol Oxidation. International Journal of Hydrogen Energy, 42 (2017) 13061–13069. https://doi.org/10.1016/j.ijhydene.2017.03.230.

[51]. Y. Yildiz, H. Pamuk, Ö. Karatepe, Z. Dasdelen, & F. Sen, Carbon Black Hybrid Material Furnished Monodisperse Platinum Nanoparticles as Highly Efficient and Reusable Electrocatalysts for Formic Acid Electro-Oxidation. RSC Advances, 6 (2016) 32858–32862. https://doi.org/10.1039/c6ra00232c.

[52]. E. Erken, Y. Yıldız, B. Kilbaş, & F. Şen, Synthesis and Characterization of Nearly Monodisperse Pt Nanoparticles for C_1 to C_3 Alcohol Oxidation and Dehydrogenation of Dimethylamine-borane (DMAB). Journal of Nanoscience and Nanotechnology, 16 (2016) 5944–5950. https://doi.org/10.1166/jnn.2016.11683.

[53]. B. Çelik, E. Erken, S. Eriş, Y. Yildiz, B. Şahin, H. Pamuk, & F. Sen, Highly monodisperse Pt(0)@AC NPs as highly efficient and reusable catalysts: The effect of The Surfactant on Their Catalytic Activities in Room Temperature Dehydrocoupling of DMAB. Catalysis Science and Technology, 6 (2016) 1685–1692. https://doi.org/10.1039/c5cy01371b.

[54]. B. Çelik, S. Kuzu, E. Erken, H. Sert, Y. Koşkun, & F. Şen, Nearly Monodisperse Carbon Nanotube Furnished Nanocatalysts as Highly Efficient and Reusable Catalyst for Dehydrocoupling of DMAB and C_1 to C_3 Alcohol Oxidation. International Journal of Hydrogen Energy, 41 (2016) 3093–3101. https://doi.org/10.1016/j.ijhydene.2015.12.138.

Materials Research Forum LLC
doi: http://dx.doi.org/10.21741/9781644900079-8

Chapter 8

Conducting Polymer-Based Microbial Fuel Cells

Ruchira Rudra[1], Prasanta Pattanayak[1], Patit Paban Kundu[2*]

[1]Advanced Polymer Laboratory, Department of Polymer Science & Technology, University of Calcutta, 92, A. P. C. Road, Kolkata - 700 009, India

[2]Department of Chemical Engineering, Indian Institute of Technology, Roorkee, Roorkee-247667, India

ppk923@yahoo.com, ppkfch@iitr.ac.in (P.P. Kundu)

Abstract

Advance technology is always incisive to alleviate the fossil fuel crisis in an eco-friendly way. A microbial fuel cell is such a sustainable, green technology that can facilitate bio-electrochemical conversion. Conducting polymers have become promising materials for high conductivity, catalytic behaviour, and excellent electrochemical activity and thus have become one of the utmost demanding materials for application in a microbial fuel cell. They have been highly recommended to modify electrodes, separators due to increased performance in terms of high conductivity, durability, power density in MFC. In this chapter, the utilization of available conducting polymer-based materials along with modification in MFC is introduced in details. Additional focus is given to future research aspirations with probable directions for future progress in this area.

Keywords

Conducting Polymer, Microbial Fuel Cell, Electrochemistry, Cathode and Anode Catalyst, Oxygen Reduction Reaction

List of Abbreviations

ACNF- Activated carbon nanofiber

AEM- Anion exchange membrane

BC- Bacterial cellulose

C- Carbon

CC- Carbon cloth

CEM- Cation exchange membrane

Materials Research Forum LLC
doi: http://dx.doi.org/10.21741/9781644900079-8

CF- Carbon felt

CNT- Carbon nanotubes

CPHs- Conductive polypyrrole hydrogels

Co- Cobalt

Fe_3O_4- Ferrous oxide

FePc- Iron phthalocyanine

G- Graphene

GF- Graphite felt

KC- Kappa-carrageenan

MFC -Microbial fuel cell

Mn- Manganese

MnO_2- Manganese oxide

$MnCo_2O_4$ NRs- Manganese cobaltite nanorods

MWCNT- Multiwalled carbon nanotubes

Ni- Nickel

NiO- Nickel oxide

NPs- Nano-particles

NT-MPMs- Multi-walled MnO_2/polypyrrole/MnO_2 nanotubes

ORR- Oxygen reduction reaction

PA- Polyacetylene

PANI/PAni- Polyaniline

PANI-LMC- Polyaniline hybridized large mesoporous carbon

PEM- Polyelectrolyte membrane

PEDOT- Poly(3,4-ethylenedioxythiophene)

PPV- Poly(p-phenylenevinylene)

PPy - Polypyrrole

PTh – Polythiophene

PU- Polyurethane

rGO- Reduced graphene oxide

SAC- Sargassum activated carbon

SPAni- Sulfonated polyaniline

SS- Stainless steel

SSFF- Stainless steel fiber felt

SS-P- Stainless steel plates

VO_2- Vanadium oxide

Materials Research Forum LLC
doi: http://dx.doi.org/10.21741/9781644900079-8

Contents

1.   Introduction

Microbial fuel cells (MFCs) belonging to a promising renewable technology are used for the conversion of electrical energy from the organic substrates. Being a useful device, MFCs utilize agricultural residue, municipal and industrial waste as a source of carbon in order to minimise the emission of greenhouse gases and global warming. Thus, MFCs are a better substitute for the naturally occurring non-renewable ubiquitous sources like fossil fuels, such as carbonaceous material (coke), crude oil, natural gas which are likely to exhaust in near future. An MFC, like other fuel cells, consists of anodic and cathodic compartments, separated by an ion permeable membrane, and are interconnected electrically through an external circuit with a load (resistor) (Fig. 1).

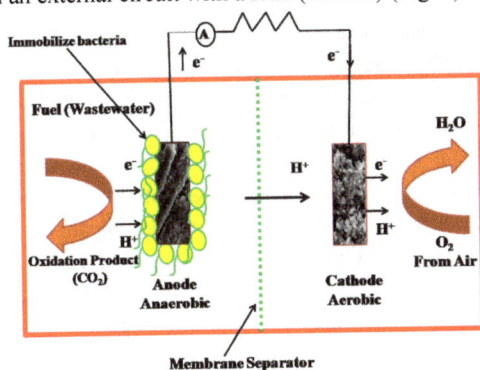

Fig. 1 Structure and components of MFC

Materials Research Forum LLC

doi: http://dx.doi.org/10.21741/9781644900079-8

In a dual chambered microbial fuel cell, bacteria, which act as biocatalyst, decompose the organic compounds to generate CO_2, proton and electron at anode under an anaerobic condition. Electrons thus produced are transferred to the cathode via an external circuit and generate a voltage between the two compartments. The protons get shifted to cathode also via polymer electrolyte membrane or salt-bridge. Fig. 2 illustrates anodic and cathodic reactions by using acetate and glucose substrate in MFC application. The performance of MFC depends on many factors like (i) inoculation of microbes, (ii) polyelectrolyte membrane (PEM), (iii) chemical substrate, (iv) external and internal resistance of cell (double and single chamber), (v) materials of the electrode and (vi) electrode spacing. Polyelectrolyte membrane (PEM) and electrode (catalyst) materials are the most significant among these factors.

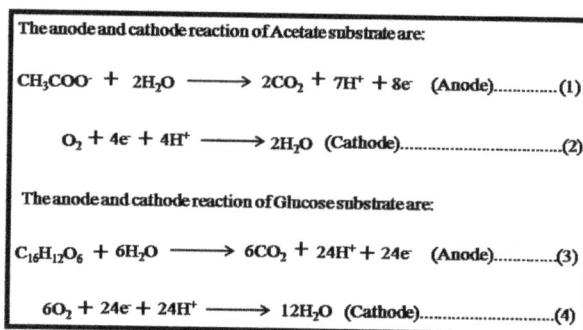

The anode and cathode reaction of Acetate substrate are:

$$CH_3COO^- + 2H_2O \longrightarrow 2CO_2 + 7H^+ + 8e^- \quad \text{(Anode)}..............(1)$$

$$O_2 + 4e^- + 4H^+ \longrightarrow 2H_2O \quad \text{(Cathode)}...........................(2)$$

The anode and cathode reaction of Glucose substrate are:

$$C_{16}H_{12}O_6 + 6H_2O \longrightarrow 6CO_2 + 24H^+ + 24e^- \quad \text{(Anode)}..............(3)$$

$$6O_2 + 24e^- + 24H^+ \longrightarrow 12H_2O \quad \text{(Cathode)}..........................(4)$$

Fig. 2 schematic representation of cathodic and anodic reactions in MFCs

By virtue of their excellent conductivity and transformable electrochemical potential, conducting polymers (CPs) are appraised as superior material for energy conversion [1]and play a vital role in increasing surface area for microbes adhesion (anode) and a large number of active sites for nucleation with air (cathode) for bioelectricity generation in MFCs. Most of the conducting polymers are aromatic in nature consisting of conjugated double bonds and contains simple elements like carbon, hydrogen, nitrogen and sulfur etc. Some of the representative CPs include polyaniline (PANI), polypyrrole (PPy), polythiophene (PTh), polyacetylene (PA), poly(3,4-ethylene dioxythiophene) (PEDOT), and poly(p-phenylenevinylene) (PPV). Fig. 3 represents the chemical structures and synthesis routes of three common CPs like polypyrrole (PPy), polyaniline (PANI), polythiophene (PTh) and a copolymer such as poly(aniline Co-pyrrole). They have outstanding capacitive properties, with conductivity ranges in between 10^{-6} Scm^{-1}

Materials Research Forum LLC
doi: http://dx.doi.org/10.21741/9781644900079-8

to 10^3 Scm^{-1}. By optimizing the doping level, their conductivities can be extended in a wide range, from 10^{-10} to 10^4 S cm^{-1}, which includes insulators, semiconductors and also conductors [2]. The chemical structures and characteristics of some common conducting polymers are shown in Table 1. For the catalyst modification process, CPs are prominent materials. The CPs and their composites have been considered most suitable matrices for the deposition of metal nanoparticles (NPs) due to their lower cost, greater specific capacitance, remarkable electrical conductivity as well as thermal stability and these can be simply synthesized by using oxidative chemical or electrochemical methods.

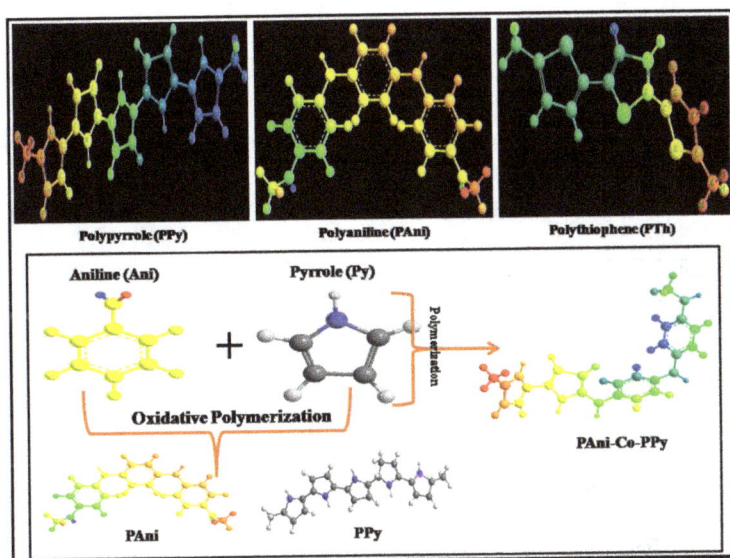

Fig. 3 chemical structures and synthesis routes of three common CPs and copolymers such as polyaniline (PAni), polypyrrole (PPy), , polythiophene (PTh), and poly(aniline Co-pyrrole)

Materials Research Forum LLC
doi: http://dx.doi.org/10.21741/9781644900079-8

Table 1: *Chemical structures and characteristics of some common conducting polymers (CPs)*

CP's Name	Chemical structure	Maximum Conductivity (Scm^{-1})	Stability
Polyaniline (PAni)		9-10	Stable
Polypyrrole (PPy)		1000-2000	Reasonably Stable
Polythiophene (PTh)		90-100	Stable
Polyacetylene		1.5×10^5	Reacts with air
Poly(p-phenylene vinylene)		1000	Stable in un-doped form

Nowadays enormous efforts are being made to utilize the application of various conducting polymers as electrodes, separators, supercapacitors and nanostructured materials in MFC.

2. Conducting polymers based membranes in MFC

Polymer electrolyte membrane plays a key role by improving the efficacy in MFC with the reduction of substrate and oxygen cross-over from the anode to cathode and vice-versa respectively. But the inherent resistance of membrane can reduce the overall performance during proton transfer mechanism. Thus increasing the conductivity of the PEM is one of the prime concerns among the researchers. Various types of separator media e.g. including proton exchange membrane (PEM), anion exchange membrane (AEM), bipolar membrane, microfiltration membrane, ultrafiltration membranes, porous

Enzymatic Fuel Cells
Materials Research Foundations **44** (2019) 173-187

Materials Research Forum LLC
doi: http://dx.doi.org/10.21741/9781644900079-8

fabrics, glass fibers, J-Cloth and salt bridge have been effectively used during the past few decades. Nafion membrane is one of the most popular membranes that has been used in MFC application due to higher conductivity and the presence of hydrophilic functional groups along with hydrophobic fluorocarbon backbone. But the higher cost, pH splitting and bio-fouling hinder its MFC application and promote to find an alternative material for improved performance. It is reported that long conjugated PANI can enhance the power density of the Nafion 112 membrane in such a way that it became comparable with Nafion 117 membrane [3]. The maximum power density and current density achieved by the use of Nafion-112/PANI composite membrane containing fuel cell are 124.03 mV m^{-2} and 454.66 mA m^{-2} respectively, which are approximately more than nine times higher than Nafion-112 membrane alone. The maximum power densities obtained from Nafion-112 and Nafion-117 membranes containing MFC are 13.9 mW and 133.3 mW respectively.

3. Conducting polymers and their composites used as catalyst support matrices in MFC

Conducting polymers act as a novel anode and cathode catalysts support material. These polymers play important role in increasing the surface area for microbes attachment (anode) and a large number of active sites for nucleation with air (cathode) for bioelectricity generation in MFCs. MFC requires highly efficient electrocatalyst that activates oxygen-reduction reaction (ORR) at the cathode compartment. It is reported that the molecular oxygen structure on CP's surface gets distorted during the chemisorptions of oxygen, for this reason, the ORR get catalyzed [4]. The selection, as well as development of proper electrode material, is an essential criterion for MFC application. Based on their performances, conducting polymers can serve as both anodes as well as cathode catalysts in MFC.

a) Conducting polymers as an anode catalyst in MFC:

The set up of electrical activity on the biofilm and the bioelectrochemical reaction take place at anode compartment. Thus, anode materials play a significant role in the overall performance of MFC. A desirable anode should have enhanced conductivity, higher specific surface area, good biocompatibility and reasonable stability. Traditionally carbon-containing materials (i.e. carbon paper, graphite sheets, carbon graphite brush, biomass bearing porous carbon and carbon cloth), stainless steel, metals such as copper, titanium, nickel, gold have been widely used as an anode electrode for MFC [5]. Among all carbon materials, graphite sheets have higher strength than carbon paper because carbon paper is very thin and relatively stiff but also slightly brittle. Alternatively, carbon cloth is the most choice because of good flexible and high porosity which provide more surface area for bacterial attachment than carbon paper. Electrode materials depend upon the physical and chemical properties like (i) chemical stability (ii) surface area and (ii) electric conductivity.

Conducting polymers such as PANI and PPy are generally used as a substitute for high-cost materials to modify the anodic performances due to their superior conductivity and lower cost. Moreover, CPs like PANI carry positive charges in the neutral ambiences which attract negatively charged bacteria for increasing in adhesion. It was reported that PANI coated anode can decrease the establishment time from 140 to 78 by enhancing bacterial adhesion [5]. Table 2 represents the performances of anodes with modification of carbon by conductive polymers in MFCs.

Table 2. The performance of anode with modification of carbon by conductive polymers in MFCs.

Modified Anode Electrode Materials	MFC Assembly	Bacteria	Power Density (mWm^{-2})	Reference
Tartaric acid doped PANI/carbon cloth	Dual chamber	S. oneidensis MR-1	490	[6]
PANI/SSFF	Dual chamber	Domestic wastewater	360	[7]
PANI/CNT/GF	Dual chamber	S. putrefaciens	257	[8]
MnFe$_2$O$_4$/PANI/CC	Single chamber	S. putrefaciens	11.2 W/m^3	[9]
PANI/CF 2016	Dual chamber	S. cerevisiae	460	[10]
Poly (3,4-ethylenedioxythiophene) (PEDOT) modified carbon cloth	Dual chamber	S. loihica strain PV-4	140	[11]
(PANI+G+CC)	single chamber	Effluent lake sediment	884	[12]
Conductive polypyrrole hydrogels and carbon Nanotubes composite (CPHs/CNTs)	Dual chamber	Mixed bacterial culture	1898	[13]

MnO$_2$/polypyrrole/MnO$_2$ nanotubes (NT-MPMs)	Single chamber	Mixed bacterial culture	934.8	[14]
BC/PANI nano-biocomposite	Dual chamber	Anaerobic sludge	616	[15]
PANI-LMC	Dual chamber	Mixed bacterial culture	1280	[16]
PPy-ACNF/CNT	Dual chamber	Shewanella oneidensis	598	[17]
PU/Graph/PPy	Dual chamber	Municipal wastewater	305.5 mW/m^3	[18]
PPy/SS	Single-chamber	Anaerobic granular sludge	1190.94	[19]
Ppy/SAC/SS	Dual chamber	Mixed bacterial culture	45.2 W/m^3	[20]
PANI/rGO	Single chamber	Mixed bacterial culture	862	[21]
SS-P/PANi	Single chamber	Synthetic wastewater	0.078 mWcm^{-2}	[22]
MWCNT-MnO$_2$/PPy	dual Chambered	Sewage waste water	1125.4	[23]

(b) Conducting polymers as cathode catalyst in MFC:

The design and fabrication of cathodic components are very critical and receiving much attention nowadays. In MFCs, different types of cathodic aqueous mediators are used as oxidants for electron acceptors such as (i) permanganate (ii) ferricyanide (iii) persulfate and (iv) dichromate for achieving higher performances. Oxygen has been considered as the most suitable electron acceptor in air cathode MFCs, because of its direct availability from the air, non-toxicity, and low cost. Thus oxygen reduction reaction (ORR) is one of the main components in MFC in order to decrease ORR overpotential. There are two steps electron transfer processes for ORR that depend on the nature of cathode catalyst. Usually, the reduction of oxygen involves two different pathways i.e. (a) 2-electron pathway and (b) 4-electron pathway (Fig. 4). Among these, the 4-eletron pathway has

Enzymatic Fuel Cells
Materials Research Foundations **44** (2019) 173-187

Materials Research Forum LLC
doi: http://dx.doi.org/10.21741/9781644900079-8

wider acceptance than the 2- electron pathways because of the 2-electron pathway produces hydrogen peroxide (H_2O_2), that causes high over potential.

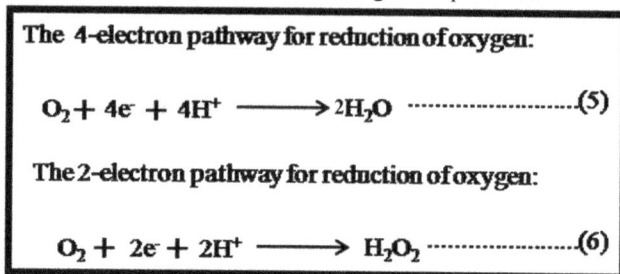

The 4-electron pathway for reduction of oxygen:

$$O_2 + 4e^- + 4H^+ \longrightarrow 2H_2O \quad \cdots\cdots\cdots\cdots(5)$$

The 2-electron pathway for reduction of oxygen:

$$O_2 + 2e^- + 2H^+ \longrightarrow H_2O_2 \quad \cdots\cdots\cdots(6)$$

Fig. 4 ORR reaction Mechanism

During the synthesis of cathode catalyst-supporting matrices, CPs have emerged as probable materials. In this regard, comparative data for their involvements in CPs as a support matrices are provided in Table 3.

Table 3. Comparative data of the MFCs performances (Power density) of the CPs as a cathode catalyst supporting matrices published by other researchers.

S. No.	Catalyst	Types of electrode used	Types of MFC used	Power density (Maximum) ($mW\ m^{-2}$)	Reference
1	Mn–PPY–CNT	Glass carbon electrode (GCE)	Air cathode-MFC	213	[24]
2	PPy/C	Non-wet proofing carbon cloth	Single chambered-MFC, air-cathode	402	[25]
3	PANI/MWNT	graphite felt	Single chamber, air cathode	488	[26]
4	Ni:Co/SPAni	Carbon Paper	Single chambered-MFC, air cathode	659.79	[27]
5	CNT/PPy nanocomposite	Carbon cloth	Dual-chamber	113.5 mW/m^2	[28]
6	PANI/C/FePc	Wet-proofed carbon cloth	Air–cathode	630.5	[29]

7	Pani–MnO_2	Carbon paper	Dual-chamber	0.0588 W m^{-2}	[30]
8	$MnCo_2O_4$ NRs/PPy	Carbon paper	Air-cathode	6.11W/m^3	[31]
9	polypyrrole (PPy)/kappa-carrageenan(KC)	Carbon paper	Dual-chamber	72.1 mW/m^2	[32]
10	rGO/PEDOT/Fe_3O_4	Carbon cloth	Single-chamber	3159 mW/m^2	[33]

4. Conducting polymer-based material as supercapacitor in MFC

The electrochemical-based capacitor i.e. supercapacitor is an electrical energy storage device consisting of cathode and anode as negative and positive electrodes respectively, along with separator. It delivers high power output through rapid charge/discharge capability. The overall performance of a supercapacitor depends on the electrode material as it is considered as the most important part in the device. Semiconducting transition metal oxides such as MnO_2, VO_2, and NiO, CuO, etc are some commonly used electrode materials that have been extensively used till date. Recently, CPs such as PANI and MnO_2 modified PPy have been utilized for improving their capacitance. Both the CPs have high conductivity, good reversibility, environmental stability, large voltage channel and low cost. Though the theoretical capacitance of PANI is 2000 F g^{-1} its electrochemical capacitance is still very low (160-815 F g^{-1}). It was reported that PANI-MnO_2 nanocomposite gained 525 F g^{-1} capacitance at a current density of 2 A g^{-1} [30]. In a similar manner, NiO@PANI-carbon felt anode material shows an excellent supercapacitive response in MFC application [34]. Based on their performance, CPs can be utilized in three different ways *viz* nanostructuring, compositing with carbon-containing nano-materials and in combination with transition metal oxides [1].

5. Current Status and challenges

Considerable progress has been made in respect of the utilization of CPs based components in MFCs application. It has been extensively observed that conducting polymer-based materials synthesized via nano-structuring, hybridization, have been suitable candidates as an electrode material (electrocatalyst) in the performance of MFC. Conducting polymer-based membranes are presently at infancy stage due to structural instability of CPs. Although CP loaded separator can efficiently prevent oxygen as well as substrate cross-over it is unable to fix pH splitting between the two chambers. This is

Materials Research Forum LLC

doi: http://dx.doi.org/10.21741/9781644900079-8

due to the transfer of limited numbers of cations and anions through the pores of the membrane, as a result, the anode chamber becomes acidified and cathode becomes alkaline in nature. This retards microbial activity, deteriorates cathode catalytic activity and thus hampered the overall cell performance get. This process is quite challenging and needs to be scaling up. Two major problems that require rectification are proton assembling within the biofilm and cathodic over potential. The electrode spacing reduces volumetric power density that causes economic losses and electrode over potential. However, a balance between the cost and overall performance of the MFC should be effectively maintained. Advanced modeling and simulation must be recommended in order to find bioelectrochemical interaction made by CPs based biofilms with the electrode.

Conclusion

CPs based MFCs are a promising technology for the production of green electricity from organic wastes. Currently, limited applications are possible due to low power generation in MFC. A deep understanding of the fundamental electrochemistry along with microbiology is essential before further advances in power output are made to happen. CPs are excellent materials with multifunctional activities in MFC application. Proper development of research and fundamental designing concept are required to build up long-term energy conversion and storage capacity in order to develop carbon-free technology in future.

References

[1]   J. Wang, J. Wang, Z. Kong, K. Lv, C. Teng, Y. Zhu, Conducting-polymer-based materials for electrochemical energy conversion and storage, Adv. Mater. 1703044 (2017) 1-11.

[2]   H. Wang, J. Lin, Z. Xiang Shen, Polyaniline (PANI) based electrode materials for energy storage and conversion, J. Sci. Adv. Mat. Devices. 1 (2016) 225-255.

[3]   N. Mokhtarian, M. Ghasemi, W. R.W. Daud, M. Ismail, G. Najafpour, J. Alam, Improvement of microbial fuel cell performance by using Nafion polyaniline composite membranes as a separator, J. Fuel Cell Sci. Technol. 10 (2013) 041008.

[4]   K. B. Liew, W.R.W. Daud, M. Ghasemi, J.X. Leong, S.S. Lim, M. Ismail, Non-Pt catalyst as oxygen reduction reaction in microbial fuel cells: A review, Int. J. Hydrogen Energ. 39 (2014) 4870 -4883.

Enzymatic Fuel Cells   Materials Research Forum LLC
Materials Research Foundations **44** (2019) 173-187   doi: http://dx.doi.org/10.21741/9781644900079-8

[5]   J. Li, W. Yang, B. Zhang, D. Ye, X. Zhu, Q. Liao, Electricity from microbial fuel
      cells, in: Q. Liao, J.-S. Chang, C. Herrmann, A. Xia (Eds.), Bioreactors for
      Microbial Biomass and Energy Conversion, 2017, pp. 391-434.

[6]   Z.H. Liao, J.Z. Sun, D.Z. Sun, R.W. Si, Y.C. Yong , Enhancement of power
      production with tartaric acid doped polyaniline nanowire network modified anode
      in microbial fuel cells, Bioresour. Technol. 192 (2015) 831-834.

[7]   J. Hou, Z. Liu, Y. Li, Polyaniline modified stainless steel fiber felt for high-
      performance microbial fuel cell anodes, J. Clean Energy Technol. 3 (2015) 165-
      169.

[8]   H.F. Cui, L. Du, P.B. Guo, B. Zhu, J.H.T. Luong, Controlled modification of
      carbon nanotubes and polyaniline on macroporous graphite felt for high-
      performance microbial fuel cell anode, J. Power Sourc. 283 (2015) 46–53.

[9]   S. Khilari, S. Pandit, J.L. Varanasi, D. Das, D. Pradhan, Bifunctional manganese
      ferrite/polyaniline hybrid as electrode material for enhanced energy recovery in
      microbial fuel cell, ACS Appl. Mater. Interf. 7 (2015) 20657–20666.

[10] D. Hidalgo, T. Tommasi, S. Bocchini, A. Chiolerio, A. Chiodoni, I. Mazzarino, B.
      Ruggeri, Surface modification of commercial carbon felt used as anode for
      microbial fuel cells, Energy, 99 (2016) 193-201.

[11] X. Liu, W. Wu, Z. Gu, Poly(3,4-ethylenedioxythiophene) promotes direct electron
      transfer at the interface between Shewanella loihica and the anode in a microbial
      fuel cell, J. Power Sourc. 277 (2015) 110-115.

[12] L. Huang, X. Li, Y. Ren, X. Wang, In-situ modified carbon cloth with
      polyaniline/graphene as anode to enhance performance of microbial fuel cell,
      Int. J. Hydrogen Energ. 41 (2016) 11369-11389.

[13] X. Tang, H. Li, Z. Du, W. Wang, H.Y. Ng, Conductive polypyrrole hydrogels and
      carbon nanotubes composite as an anode for microbial fuel cells, RSC Adv. 5
      (2015) 50968-50974.

[14] H. Yuan, L. Deng, Y. Chen, Y. Yuan, MnO_2/Polypyrrole/MnO_2 multi-walled-
      nanotube-modified anode for high-performance microbial fuel cells,
      Electrochim. Acta 196 (2016) 280-285.

[15] M. Mashkour, M. Rahimnejad, M. Mashkour, Bacterial cellulose-polyaniline
      nano-biocomposite: a porous media hydrogel bioanode enhancing the performance
      of microbial fuel cell, J. Power Sourc. 325 (2016) 322-328.

[16]   L. Zou, Y. Qiao, C. Zhong, C. M. Li, Enabling fast electron transfer through both bacterial outer-membrane redox centers and endogenous electron mediators by polyaniline hybridized large-mesoporous carbon anode for high-performance microbial fuel cells, Electrochim. Acta. 229 (2017) 31-38.

[17]   H.Y. Jung, S.H. Roh, Carbon nanofiber/polypyrrole nanocomposite as anode material in microbial fuel cells, J. Nanosci. Nanotechnol. 17 (2017) 5830-5833.

[18]   P. Pérez-Rodríguez, V.M. Ovando-Medina, S.Y. Martínez-Amador, J.A. Rodríguez-de la Garza, Bioanode of polyurethane/graphite/polypyrrole composite in microbial fuel cells, Biotechnol. Bioprocess Eng. 21 (2016) 305-313.

[19]   K. Pu, Q. Ma, W. Cai, Q. Chen, Y. Wang, F. Li, Polypyrrole modified stainless steel as high performance anode of microbial fuel cell, Biochem. Eng. J. 132 (2018) 255-261.

[20]   G. Wu, H. Bao, Z. Xia, B. Yang, L. Lei, Z. Li, C. Liu, Polypyrrole/sargassum activated carbon modified stainless-steel sponge as high-performance and low-cost bioanode for microbial fuel cells, J. Power Sourc. 384 (2018) 86-92.

[21]   N. Zhao, Z. Ma, H. Song, D. Wang, Y. Xie, Polyaniline/reduced graphene oxide-modified carbon fiber brush anode for high-performance microbial fuel cells, Int. J. Hydrog. Energ. 43 (2018) 17867-17872.

[22]   J.M. Sonawane, S. Al-Saadi, R.K.S. Raman, P.C. Ghosh, S.B. Adeloju, Exploring the use of polyaniline-modified stainless steel plates as low-cost, high-performance anodes for microbial fuel cells, Electrochim. Acta. 268 (2018) 484-493.

[23]   P. Mishra, R. Jain, Electrochemical deposition of MWCNT-MnO_2/PPy nano-composite application for microbial fuel cells, Int. J. Hydrog. Energ. 41 (2016) 22394-22405.

[24]   M. Lu, L. Guo, S. Kharkwal, H.Y. Ng, S.F.Y. Li, Manganese–polypyrrole–carbon nanotube, a new oxygen reduction catalyst for air-cathode microbial fuel cells, J. Power Sources, 221 (2013) 381-386.

[25]   Y. Yuan, S. Zhou, L. Zhuang, Polypyrrole/carbon black composite as a novel oxygen reduction catalyst for microbial fuel cells, J. Power Sources, 195 (2010) 3490-3493.

[26]   H. Cui, L. Du, P. Guo, B. Zhu, J.H. Luong, Controlled modification of carbon nanotubes and polyaniline on macroporous graphite felt for high-performance microbial fuel cell anode, J. Power Sources, 283 (2015) 46-53.

Enzymatic Fuel Cells
Materials Research Foundations **44** (2019) 173-187

Materials Research Forum LLC
doi: http://dx.doi.org/10.21741/9781644900079-8

[27] F. Papiya, P. Pattanayak, P. Kumar, V. Kumar, P.P. Kundu, Development of highly efficient bimetallic nanocomposite cathode catalyst, composed of Ni: Co supported sulfonated polyaniline for application in microbial fuel cells, Electrochim. Acta. 282 (2018) 931-945.

[28] M. Ghasemi, W.R.W. Daud, S.H.A. Hassan, T. Jafary, M. Rahimnejad, A. Ahmad, M.H. Yazdi, Carbon nanotube/polypyrrole nanocomposite as a novel cathode catalyst and proper alternative for Pt in microbial fuel cell, Int. J. Hydrog. Energ. 41 (2016) 4872-4878.

[29] Y. Yuan, J. Ahmed, S. Kim, Polyaniline/carbon black composite-supported iron phthalocyanine as an oxygen reduction catalyst for microbial fuel cells, J. Power Sources, 196 (2011) 1103-1106.

[30] S.A. Ansari, N. Parveen, T.H. Han, M.O. Ansari, M.H. Cho, Fibrous polyaniline@ manganese oxide nanocomposites as supercapacitor electrode materials and cathode catalysts for improved power production in microbial fuel cells, Phys. Chem. Chem. Phys. 18 (2016) 9053-9060.

[31] S. Khilari, S. Pandit, D. Das, D. Pradhan, Manganese cobaltite/polypyrrole nanocomposite-based air-cathode for sustainable power generation in the single-chambered microbial fuel cells, Biosens. Bioelectron. 54 (2014) 534-540.

[32] C. Esmaeili, M. Ghasemi, L.Y. Heng, S.H.A. Hassan, M.M. Abdi, W.R.W. Daud, H. Ilbeygi, A.F. Ismail, Synthesis and application of polypyrrole/carrageenan nano-bio composite as a cathode catalyst in microbial fuel cells, Carbohydr. Polym. 114 (2014) 253-259.

[33] C.J. Kirubaharan, D.J. Yoo, A.R. Kim, Graphene/poly(3, 4-ethylenedioxythiophene)/Fe$_3$O$_4$ nanocomposite–An efficient oxygen reduction catalyst for the continuous electricity production from wastewater treatment microbial fuel cells, Int. J. Hydrog. Energ. 41 (2016) 13208-13219.

[34] D. Zhong, X. Liao, Y. Liu, N. Zhong, Y. Xu, Enhanced electricity generation performance and dye wastewater degradation of microbial fuel cell by using a petaline NiO@ polyaniline-carbon felt anode, Bioresour. Technol. 258 (2018) 125-134.

Keyword Index

About the Editors

Dr. Inamuddin is currently working as Assistant Professor in the Chemistry Department, Faculty of Science, King Abdulaziz University, Jeddah, Saudi Arabia. He is a permanent faculty member (Assistant Professor) at the Department of Applied Chemistry, Aligarh Muslim University, Aligarh, India. He obtained the Master of Science degree in Organic Chemistry from Chaudhary Charan Singh (CCS) University, Meerut, India, in 2002. He received his Master of Philosophy and Doctor of Philosophy degrees in Applied Chemistry from Aligarh Muslim University (AMU), India, in 2004 and 2007, respectively. He has extensive research experience in multidisciplinary fields of Analytical Chemistry, Materials Chemistry, and Electrochemistry and, more specifically, Renewable Energy and Environment. He has worked on different research projects as project fellow and senior research fellow funded by the University Grants Commission (UGC), Government of India, and the Council of Scientific and Industrial Research (CSIR), Government of India. He has received the Fast Track Young Scientist Award from the Department of Science and Technology, India, to work in the area of bending actuators and artificial muscles. He has completed four major research projects sanctioned by the University Grant Commission, Department of Science and Technology, Council of Scientific and Industrial Research, and Council of Science and Technology, India. He has published 133 research articles in international journals of repute and eighteen book chapters in knowledge-based book editions published by renowned international publishers. He has published forty two edited books with Springer, United Kingdom, Elsevier, Nova Science Publishers, Inc. U.S.A., CRC Press Taylor & Francis Asia Pacific, Trans Tech Publications Ltd., Switzerland and Materials Research Forum LLC, U.S.A. He is the member of various editorial boards of journals and is serving as associate editor for journals such as Environmental Chemistry Letter, Applied Water Science, Euro-Mediterranean Journal for Environmental Integration, Springer-Nature, Frontiers Section Editor of Current Analytical Chemistry, published by Bentham Science Publishers, editorial board member for Scientific Reports-Nature and editor for Eurasian Journal of Analytical Chemistry. He has attended as well as chaired sessions in various international and national conferences. He has worked as a Postdoctoral Fellow, leading a research team at the Creative Research Initiative Center for Bio-Artificial Muscle, Hanyang University, South Korea, in the field of renewable energy, especially biofuel cells. He has also worked as a Postdoctoral Fellow at the Center of Research Excellence in Renewable Energy, King Fahd University of Petroleum and Minerals, Saudi Arabia, in the field of polymer electrolyte membrane fuel cells and computational fluid dynamics of polymer electrolyte membrane fuel cells. He is a life member of the Journal of the Indian

Chemical Society. His research interest includes ion exchange materials, a sensor for heavy metal ions, biofuel cells, supercapacitors and bending actuators.

Dr. Mohammad Faraz Ahmer is presently working as Assistant Professor in the Department of Electrical Engineering, Mewat Engineering College, Nuh Haryana, India, since 2012 after working as Guest Faculty in University Polytechnic, Aligarh Muslim University Aligarh, India, during 2009-2011. He completed M.Tech. (2009) and Bachelor of Engineering (2007) degrees in Electrical Engineering from the Aligarh Muslim University, Aligarh in the first division. He obtained a Ph.D. degree in 2016 on his thesis entitled "Studies on Electrochemical Capacitor Electrodes". He has published six research papers in reputed scientific journals. He has edited two books with Materials Research Forum LLC, U.S.A. His scientific interests include electrospun nano-composites and supercapacitors. He has presented his work at several conferences. He is actively engaged in searching of new methodologies involving the development of organic composite materials for energy storage systems.

Mohd Imran Ahamed is a Research Scholar at the Department of Chemistry, Aligarh Muslim University, Aligarh, India. He is working towards his Ph.D. thesis entitled Synthesis and characterization of inorganic-organic composite heavy metals selective cation-exchangers and their analytical applications. He has published several research and review articles in journals of international recognition. He has also edited three books published by Springer and Materials Research Forum LLC, U.S.A. He has completed his Bachelor of Science (Chemistry) at the Aligarh Muslim University, Aligarh, India, and Masters in Chemistry (Organic Chemistry) at the Dr. Bhimrao Ambedkar University, Agra, India. His research work includes ion exchange chromatography, wastewater treatment, and analysis, bending actuator and electrospinning.

Prof. Abdullah M. Asiri is the Head of the Chemistry Department at the King Abdulaziz University since October 2009 and he is the founder and the Director of the Center of Excellence for Advanced Materials Research (CEAMR) since 2010 till date. He is the Professor of Organic Photochemistry. He graduated from King Abdulaziz University (KAU) with B.Sc. in Chemistry in 1990 and received his Ph.D. from the University of Wales, College of Cardiff, U.K. in 1995. His research interest covers color chemistry, synthesis of novel photochromic and thermochromic systems, synthesis of novel coloring matters and dyeing of textiles, materials chemistry, nanochemistry and nanotechnology, polymers and plastics. Prof. Asiri is the principal supervisors of more than 20 M.Sc. and six Ph.D. theses. He is the main author of ten books of different chemistry disciplines. Prof. Asiri is the Editor-in-Chief of King Abdulaziz University Journal of Science. A major achievement of Prof. Asiri is the discovery of tribochromic compounds, a new class of compounds which change from slightly or colorless to deep colored when subjected to small pressure or when grind. This discovery was introduced to the scientific community as a new terminology published by IUPAC in 2000. This discovery was awarded a patent from the European Patent office and from the UK patent. Prof. Asiri is involved in many committees at the KAU level and on the national level. He took a major role in the advanced materials committee working for KACST to identify the national plan for science and technology in 2007. Prof. Asiri played a major role in advancing the chemistry education and research in KAU. He has been awarded the best researchers from KAU for the past five years. He also awarded the Young Scientist Award from the Saudi Chemical Society in 2009 and also the first prize for the distinction in science from the Saudi Chemical Society in 2012. He futher received a recognition certificate from the American Chemical Society (Gulf region Chapter) for the advancement of chemical science in the Kingdome. He received a Scopus certificate for the most publishing scientist in Saudi Arabia in chemistry in 2008. He is a member of the editorial board of various journals of international repute. He is the Vice- President of the Saudi Chemical Society (Western Province Branch). He holds four USA patents, more than one thousand publications in international journals, several book chapters and edited books.

www.ingramcontent.com/pod-product-compliance
Lightning Source LLC
Chambersburg PA
CBHW052013230326

41598CB00078B/3212